大连理工大学出版社

HIGHLIGHT
COMMERCIAL SPACE

商业空间

Design & Vision 工作室 编

大连理工大学出版社
Dalian University of Technology Press

图书在版编目(CIP)数据

商业空间 : 汉、英 / Design& Vision工作室编. —
大连 : 大连理工大学出版社, 2011.3
ISBN 978-7-5611-6022-0

Ⅰ. ①商… Ⅱ. ①D… Ⅲ. ①商业—服务建筑—室内
设计—作品集—世界—现代—汉、英 Ⅳ. ①TU247

中国版本图书馆CIP数据核字（2011）第017092号

出版发行：大连理工大学出版社
　　　　　（地址：大连市软件园路80号　邮编：116023）
印　　刷：利丰雅高印刷（深圳）有限公司
幅面尺寸：210mm×285mm
印　　张：19
插　　页：4
出版时间：2011年3月第1版
印刷时间：2011年3月第1次印刷
责任编辑：初　蕾
责任校对：仲　仁
装帧设计：连　帅

ISBN 978-7-5611-6022-0
定　　价：248.00元

电　话：0411-84708842
传　真：0411-84701466
邮　购：0411-84703636
E-mail：designbook@yahoo.cn
URL：http:// www.dutp.cn

如有质量问题请联系出版中心：（0411）84709043 84709246

HIGHLIGHT
COMMERCIAL SPACE

CONTENTS

RETAIL

RESTAURANT

HOTEL&OFFICE

RETAIL

Designer: Jaime Hayon
Client: Faberge
Photography: Nienke Klunder
Total area: 250 m²

Faberge Salon

The boutique expresses a world of luxury, importance and elegance as originally but through a design language that is more approachable, lightly enjoyable, and bright and unbelievably etiquette. There is a great presence of intimidation as you are approaching the full height interconnecting doors with the tinted glass of cut out gem-shaped panels. Through this gem panel pattern designer has successfully retained natural day light coming through the interior and glorifying it.

The interior has a very different feel. Not intimidating once you are inside, but inviting and jewel like sparkling. All finishes are the most elite and most sought after materials such as Carrara marble, rare woods and silk wall drapes. These finishes in combination with the highest skill in craftsmanship and an excellence in detailing are what make the experience what it is.

The furniture is typical of Hayon's signature with wooden finishes that are quite precious. The same applies for all lighting fixtures. The ceiling pendants follow this clean modern cut design with a hint of old world and glamour through their metallic shine. The silvered metallic shades express the luxurious element and retain the aspect of the organic design. These features contribute to the boutique's old meets mod and mod respects old.

The luxury is successfully incorporated in the space by the mirrored ceiling. Nothing is more high jewellery and erotic than a mirrored ceiling. The idea that all these exquisite finishes are interpreted twice makes you ecstatic and allured. Visualize a grand dame of Chanel origin with her black hat of feathers trying on a glorified diamante ring and reflecting its unique cut on the ceiling. This is exactly how these jewels are meant to be adored.

设计师以平和的、轻盈的、令人愉快的设计语汇营造了奢华、雅致的氛围。入口处通高的大门上装饰着宝石形状的开洞，其中镶嵌的彩色玻璃使室内的自然光线透射出来。内部空间所用的装饰材料十分精致，包括白色大理石、稀有的木材和丝制的帷幔。这些材料搭配上高级手工工艺和细部设计，迎合了珠宝熠熠生辉的本质。使用木制家具是这位设计师的鲜明特色，与之相对的是银色的金属灯具和突显奢华感的镜面天花。通过这些元素，设计师将传统与现代的对比完美融合在空间中。

Designer: Chikara Ohno \ sinato
Client: DURAS Inc.
Photography: Takumi Ota
Total area: 126.76 ㎡

DURAS Daiba

. . . . The long way going at floor level.

. . . . The shortcut going up some steps.

plan

At first, we have to think how to make use of the 3.65m high-ceiling for this interior. Generally the space above head height is just void for only looking in a boutique, because most of the action for buying and selling is centered close to human body. To avoid this condition, we installed imaginary ceiling made by expand-metal at 2.25m high and set up stepped platform that allows shoppers to reach the attic.

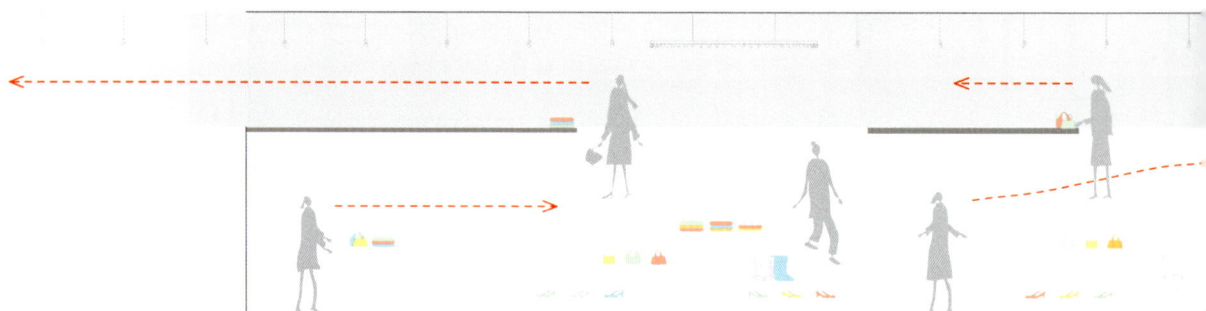

mirror image ←——→ real image

Two stepped platforms as like hills are useful stage for displaying bags, heels and mannequins and give us the choices of flow, the long way by going at floor level or the shortcut by going up some steps. Mirrors round the edge of the attic visually expand the shop. We expect this illusionary view as a gimmick not to get bored with the shopping in a huge mall.

3.65米的层高是这一项目带给设计师的挑战。对于精品店来说，因为大多数购买行为发生在符合人体高度的范围内，所以高于头顶的空间通常是无效用的。为了打破这种常规，设计师在2.25米的地方增设了天花，顾客可以通过台阶式的平台到达那里。两个台阶式的平台被用为展示皮包、高跟鞋和服饰的舞台。夹层四周的镜面从视觉上延展了空间，带给顾客不同于大型购物场所的惊喜感受。

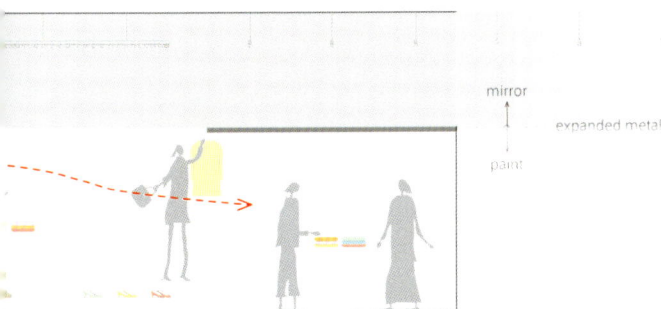

mirror

expanded metal

paint

diagram

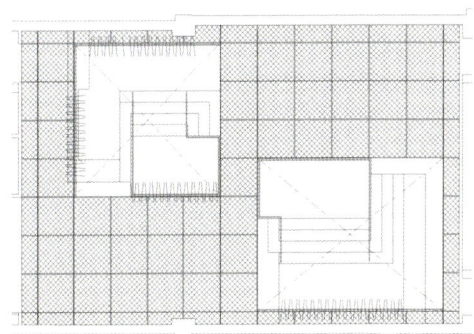

upper level plan

Designer:Chikara Ohno\ sinato
Client: Diesel Denim Gallery Aoyama
Curator: Masaaki Takahashi (www.brizhead.jp)
Photography: Toshiyuki Yano (info@yanofoto.com)

Rolls

DIESEL DENIM GALLERY AOYAMA has high reputation by curating and featuring a lot of talented creators like a video director Timothy Saccenti (Partizan) and an architect Makoto Tanijiri. Among other exhibitors are an architect Kimihiko Okada whose work Another Geography was scaled up at Museum of Contemporary Art Tokyo and an artist Mark Jenkins who was chosen by Time magazine as one of Top Ten Guerrilla Artists. Now the gallery is known as a gateway to success for new artists and architects.

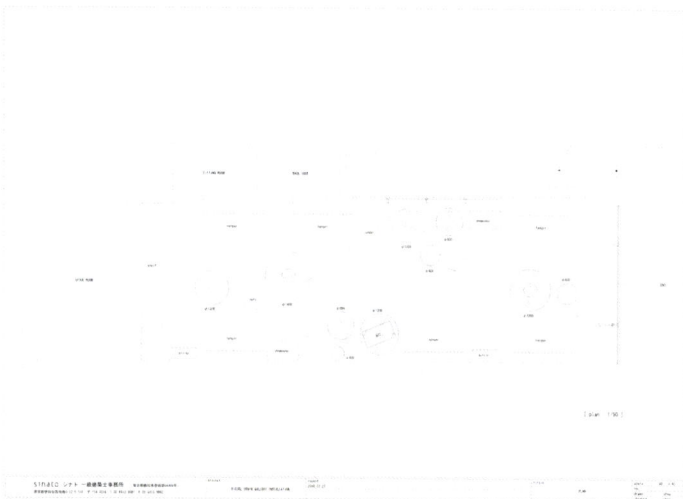

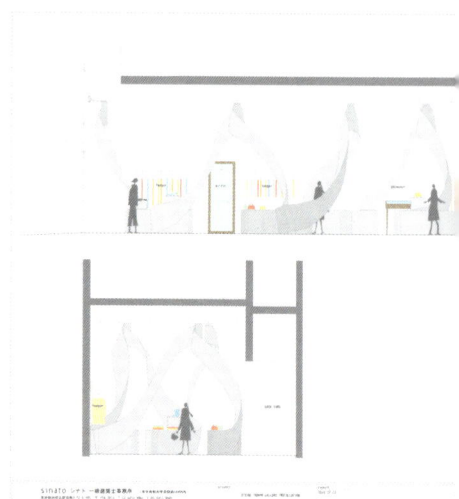

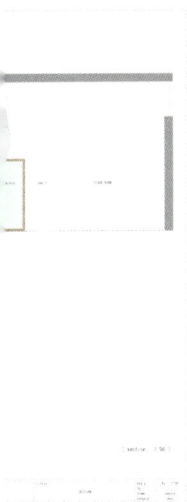

The characteristic of the material used for this installation, which is aluminum, is that it is very thin and easily bent by hands, yet harder than cloth or paper. Therefore it possesses both soft and hard qualities. By winding and sometimes extending this single, long strip of aluminum from the entrance to the back-end of the store, it creates a beautiful waving form, changing its function and features as the material strength changes. This flexible quality of the material represents a gentle connection between the softness of clothes and hardness of architecture.

位于日本青山的Diesel Denim Gallery以策划展览和推荐富有天赋的创意人而闻名于世。如今，这家美术馆已被新锐艺术家和建筑师们视为通向成功的大门。本项目设计中所使用的铝材，具有非常薄和易于用手弯折的特性，但它们又比布料或纸张更结实，因此将柔软与硬朗完美融合。呈长条状的铝材从商店的入口一直延伸到后部空间，营造了优美的波浪造型，改变了自身的功能。这种材料柔韧的特性将服饰的柔软和建筑的硬朗体现得淋漓尽致。

Designer: Gwenael Nicolas \ CURIOSITY
Contractor: Ueno Denki
Photography: Nacasa & Partners
Location: Tokyo

LONGCHAMP Flash Shop

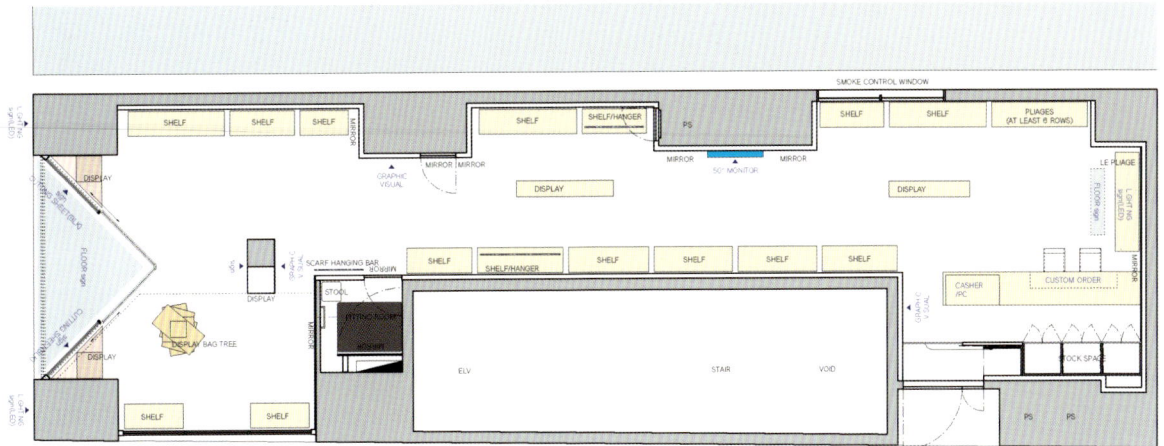

LONGCHAMP Flash Shop is an attempt to create a different paradigm for retail. It is not designed as a space but as a sequence in the busy landscape of Tokyo. A sequence should tell a story in a nano second. The flash of a camera always attracts the attention even of the most indifferent person, so it was the perfect connection with the fashionable and ludique message of the collection. The 'flash shop' was born.

A stroboscope was installed behind the new glass facade. The shadow of the black facade suddenly reveals the zebra graphic and the LONGCHAMP logo. The image already disappeared though the image is still printed in the retina of the viewer. The facade is invisible though it appears like magic, is this retail or fairy tale?

The shop layout is also crucial to reveal itself in the eyes of the person walking down the street. The front window is placed at 45 degree, perpendicular to the view of the customers, so as you walk by the display is facing you. As you walk closer the sliding doors open widely at an angle. The interior elements are reduced to the utmost minimum as products are placed on extra thin aluminum shelves suspended from semi-transparent layers of fabrics. It is like walking through floating products that appear behind white veils.

设计师希望借此设计为零售店设定一种不同的模式。这不仅仅是繁忙的东京街头的一个空间，而是街景中的一个镜头，可以在转瞬间传递一个故事。照相机的闪光灯通常会吸引大多数人的注意，因此这是传递时尚信息的完美途径。黑色玻璃立面后面安装了一个频闪观测仪，斑马纹和品牌标识会突然出现，在图像仍然保留在观者视网膜上的时候，图像已经消失了。立面看似一张图片，但又是不可见的；橱窗按照45度角排列，与人们的视角垂直；而室内装饰元素被缩减到最少，这是零售店还是一个童话呢？

LONGCHAMP

Designer: Anagrama
Collaboration: Roberto Treviño & German Deheza
Location: San Pedro, Mexico

Theurel & Thomas

Theurel & Thomas is the first pâtisserie in Mexico specialized in French macarons, the most popular dessert of the French pastries. For this project it was very important to create an imposing brand that would emphasize the unique value, elegance and detail of this delicate dessert.

One of the most important extensions of a brand, which has a business based in store selling, is the design and ambiance of the stores. The pâtisserie of Theurel & Thomas has an enlighten space with an exclusivity and elegance atmosphere. The store location is found in San Pedro, Mexico. Latin America's most affluent suburb.

NO.350 I L.17

LUNES A SÁBADO
10:00 AM - 8:00 PM

White is a central part of the design and it plays as a contrast with the colors of the French macarons. Details were an essential part of the designer's work. They meticulously selected each porcelain piece making a balance with sophisticated specks that made the value of the brand and the exclusivity of the product outshine.

Theurel & Thomas是墨西哥第一家专营法式甜点马卡龙（macaron）的甜点店，因此对于这个项目来说，营造一种令人难忘的室内空间用于突显这款精致甜点的价值、雅致和细节是非常重要的。项目所在地位于墨西哥的San Pedro，那里是拉丁美洲最富足的城郊地区。设计师以白色为商店空间的主色调，这与马卡龙甜点的颜色形成了鲜明的对照。设计师对细节的关注还体现在精心挑选的瓷质配饰上，它们与精致的产品相映成趣。

Catering

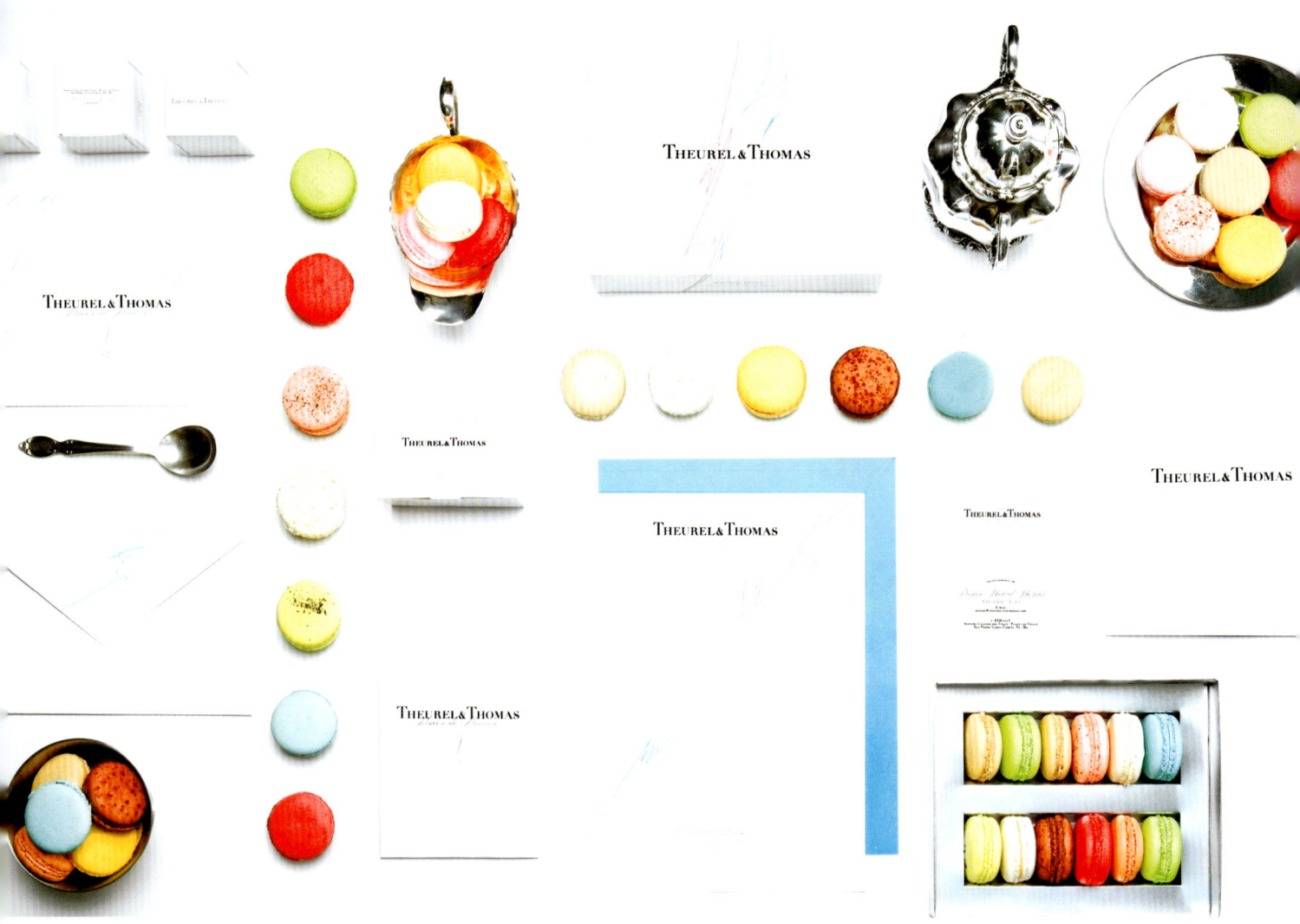

THEUREL & THOMAS

Designer: Rafael de Cárdenas
Location: Greenwich Village, New York
Photography: Floto + Warner

OHWOW Book Club

Drawing inspiration from stepping patterns commonly found in Navajo blankets, Rafael de Cárdenas constructs a new retail space for OHWOW, the creative collective spearheaded by Al Moran and Aron Bondaroff. OHWOW Book Club is located below street level in a landmarked historic brownstone on Waverly Place. At 150 sq ft, this pocket-sized store was conceived by de Cárdenas to echo a classic black & white, pre-war NYC bathroom.

Its shelving units appear stacked one atop the other and the negative space behind the shelves lends a floating sensation. A layered pattern of stream-lined brushstrokes on the walls, coupled with reflective angular mylar shapes and sharp fluorescent lighting give the space a sense of disorientation and chaos, fitting in OHWOW's vision of creating a heterotopic arena for cultural projects.

OHWOW书吧位于Waverly Place一处有历史意义的赤褐色砂石建筑中。设计师从纳瓦霍毛毯上常见的阶梯状图案上获取灵感，为这间14平方米的书吧营造了一个全新的零售空间。书架是一个堆叠在另一个之上，其后面的负空间借用了一种浮动的观感。墙面上层状的流线型图案、反光的聚酯薄膜及荧光灯给人一种迷惑和混乱的感觉，而这恰恰与OHWOW书吧所推崇的文化项目不谋而合。

Designer: Jaime Hayon
Client: Octium
Photography: Nienke Klunder
Total area: 200 ㎡

Octium Jewelry

Octium is a luxury jewelry shop that features the work of a very exclusive selection of artists and designers. Hayon's design for this project is a new approach to interior design: very different to what the luxury world has seen before. Tradition, modernity, the use of noble materials and technology are all harmoniously combined in this fantastical space.

One of the most important and innovative characteristics of this project is the division of areas and the ways of displaying the jewelry. There are many different approaches and ways to visualize and offer jewelry to the client. This creates a strong appeal for the viewer as the journey through the interior is anything but monotonous.

After passing through a very masculine facade in St. Laurent dark marble and an impressive brass door, one enters a surprisingly contrasting interior that is completely organic and very soft, like a woman's skin. The use of feminine shapes is a constant all throughout the project. Inspired on traditional Mediterranean constructions, where lime was used for finishes to give an organic feel to interiors, there are no corners inside the shop.

Innovation and detail are key aspects to the project's conception. From a very theatrical large furniture where ceramic lamps fall inside the display elements like act of magic, to a 7 meter centipede-like display element made in natural walnut and brass crutch like legs that honor Dali's influence.

There is a striking central area with a very complex multi-legged rounded table with springing cylinders that exhibit jewelry and a light installation on top of this element where cylinders of different diameters hang from the ceiling like stalactites.

Everything from furniture, chair finishes, lamps, handles, was custom-designed and made specially for Octium. The use of contrasting finishes like glossy lacquered woods, natural walnut, shiny ceramics and luxurious fabrics, give a general feel of tradition and technology joined in a balanced proportion. Octium is yet another one of these special Hayon interiors where fantasy and imagination meet functionality and quality.

Octium是一家销售艺术家高端设计系列的珠宝店。设计师采用了不同于传统奢侈品店的全新的室内设计手法，将传统、现代、材料和技术完美融合。空间规划和珠宝产品的多种展示方式是这一项目最具创新的地方，使顾客在室内穿行时不会感到枯燥无味。传统的地中海式风格带来了设计灵感，无尖角的设计使室内空间充盈着一种女性的柔美感觉。戏剧化的大型展柜、7米长的胡桃木柜台、铜制的支架、似钟乳石一般的照明装置，甚至把手都是为这一项目量身定制的。不同材料的对比，如自然的胡桃木、光滑的陶瓷、奢华的布料体现了设计师将想像力与功能性和高品质相结合的功力。

Designer: Design Research Studio
Client: Joseph
Location: Old Bond Street, London
Photography: Leon Chew

Joseph's Flagship Store

The Design Research Studio team have taken an especially irreverent approach to the interior design of their given retail destination in the heart of Old Bond Street. In a manner that is reflective of the nonconformist attitude that characterizes Tom Dixon's method of design and also articulates the spirit of departure that is Joseph's new venture, the interior of the store is boldly industrial and purposefully in contrast to the slick, over-manufactured interiors of the stores surrounding it.

'Joseph allowed us to create something unique that we wanted to do, to go in and be very not Bond Street. I think that the minimal aspect that Joseph has used is still very legitimate, but it is something that has been copied widely. What we tried to do was maintain the monochrome aspect but to bring in texture. The shop is now full of really quite intense texture and relief.'

Whilst the interior is a marked departure from Joseph's past retail style, Design Research Studio has taken the muted monochrome palette that is recognizably Joseph and rearticulated it in unexpected materials, textures and finishes. First, the qualities of naturally raw materials are revealed; rough plaster inspired by traditional wattle and daub covers the walls, unpolished stone, delivered directly from the quarry is employed for fixtures.

Natural inspirations converge with the brawn of the industrial as metal ceiling covers and balustrades become features and pro tools; projectors and spot lighting more usually used for the stage, light the space. Metallic finishes and mechanized clothes rails provide more features. The result is an entirely unexpected addition to London's most refined and elegant fashion district.

伦敦邦德街上名店林立，由Design Research Studio设计的Joseph旗舰店独树一帜，其营造的明显的工业化氛围与周围过于精致的店铺形成了鲜明的对比，而这恰恰体现了设计师的设计理念和Joseph品牌的新的发展方向。一些出人预料的材料、纹理和饰面被直接用在室内空间中，如未加工的天然材料、毛面墙和用于陈列商品的未雕琢的石块等。它们与体现工业化特质的设计，如金属天棚、栏杆、衣架和通常用在舞台上的投影仪和聚光灯完美融合。这一整体方案既保留了Joseph品牌的独有元素，又为其注入了新的亮点。

Designer: Julie Schmidt-Nielsen, Marc Jay, Jenny Selldén,
Linda Vendsalu, Lawrence Mahadoo
Client: T-Magi
Photography: Enok Holsegaard
Total area: 60 m²

T-magi Tea Shop

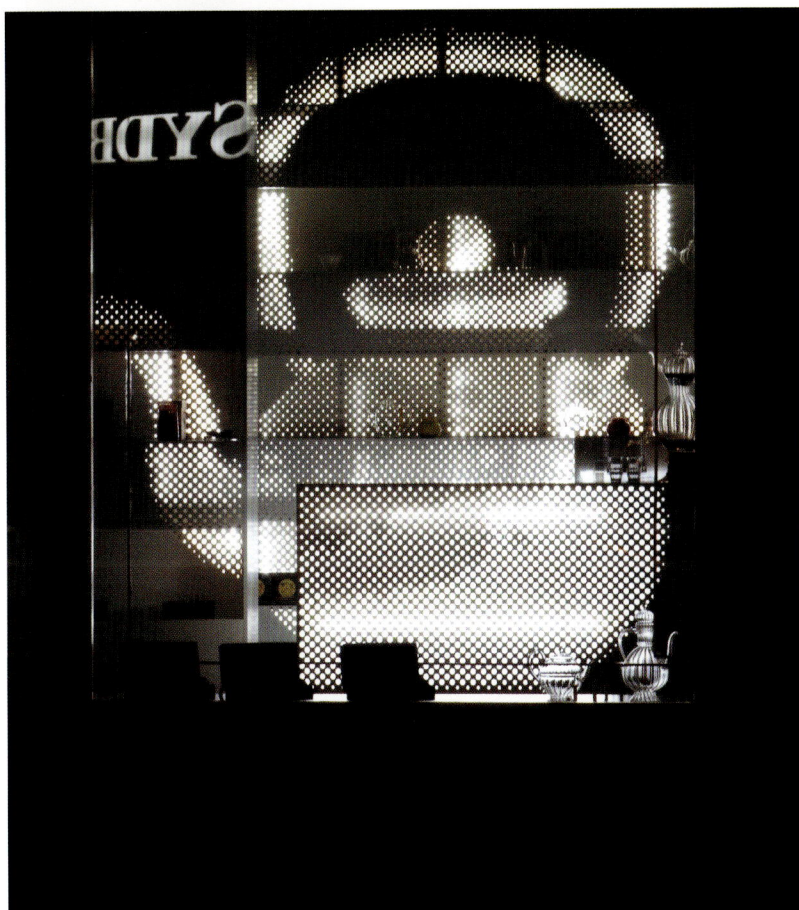

While many shops tend to have a clear distinction between storefront and interior, the design of T-magi is intended to allow the shop itself to be perceived as the display window. We have used the teapot—an object universally associated with tea—as the motif for both the shop, logo and PR material. We've intentionally designed the shop to have a lab feel to it, inviting people to pass by, smell the 40 different teas displayed on the scent wall, have a taste, read the WE-tea folder or simply browse the exclusive products of Mariage Frere sold in the shop. Tiny backlit holes perforate the shelves and back wall providing a large 3-dimensional image of a teapot; a powerful eye-catcher. The image of the teapot dissolves as you approach and becomes the furnishings of the shop.

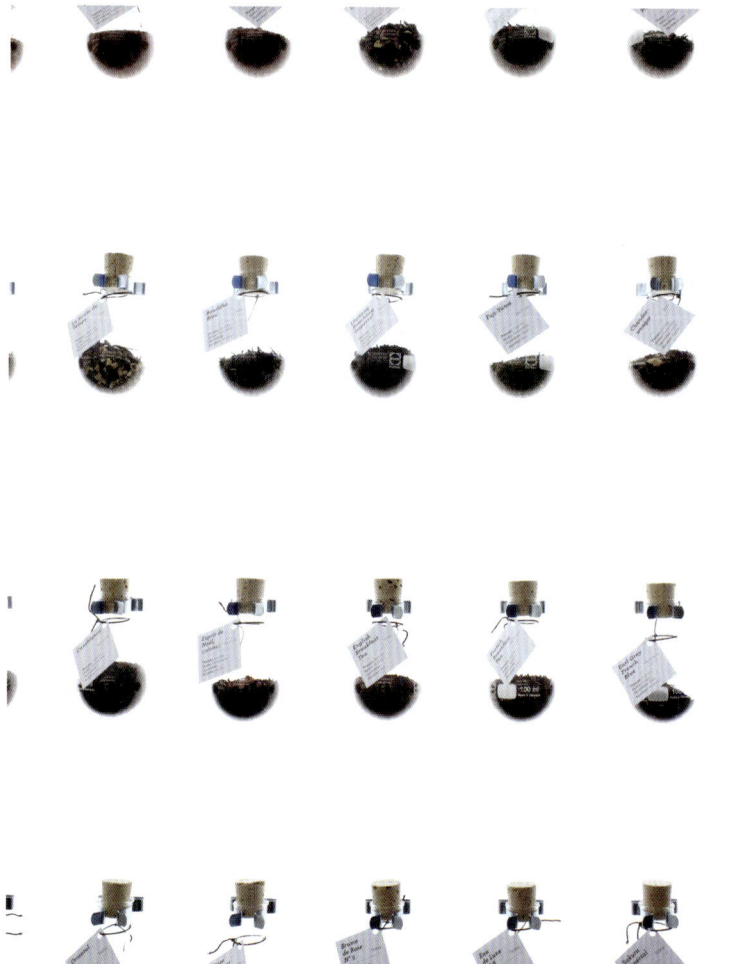

T-magi商店的设计师将商店本身与橱窗展示的功能等同起来，这与大多数商店将门面与室内空间清晰地分隔开是不同的。设计师采用与茶叶有内在关联的茶壶作为设计的装饰图案，有意使室内空间具有一种试验的氛围——邀请路过的人品尝展示在墙面上的40种不同品种的茶叶，阅读资料，或者仅仅是浏览一下商店里销售的独特的商品。展架和背板上的孔洞组合成了3D效果的茶壶的图案，这是一个有力的元素，与空间完美融合。

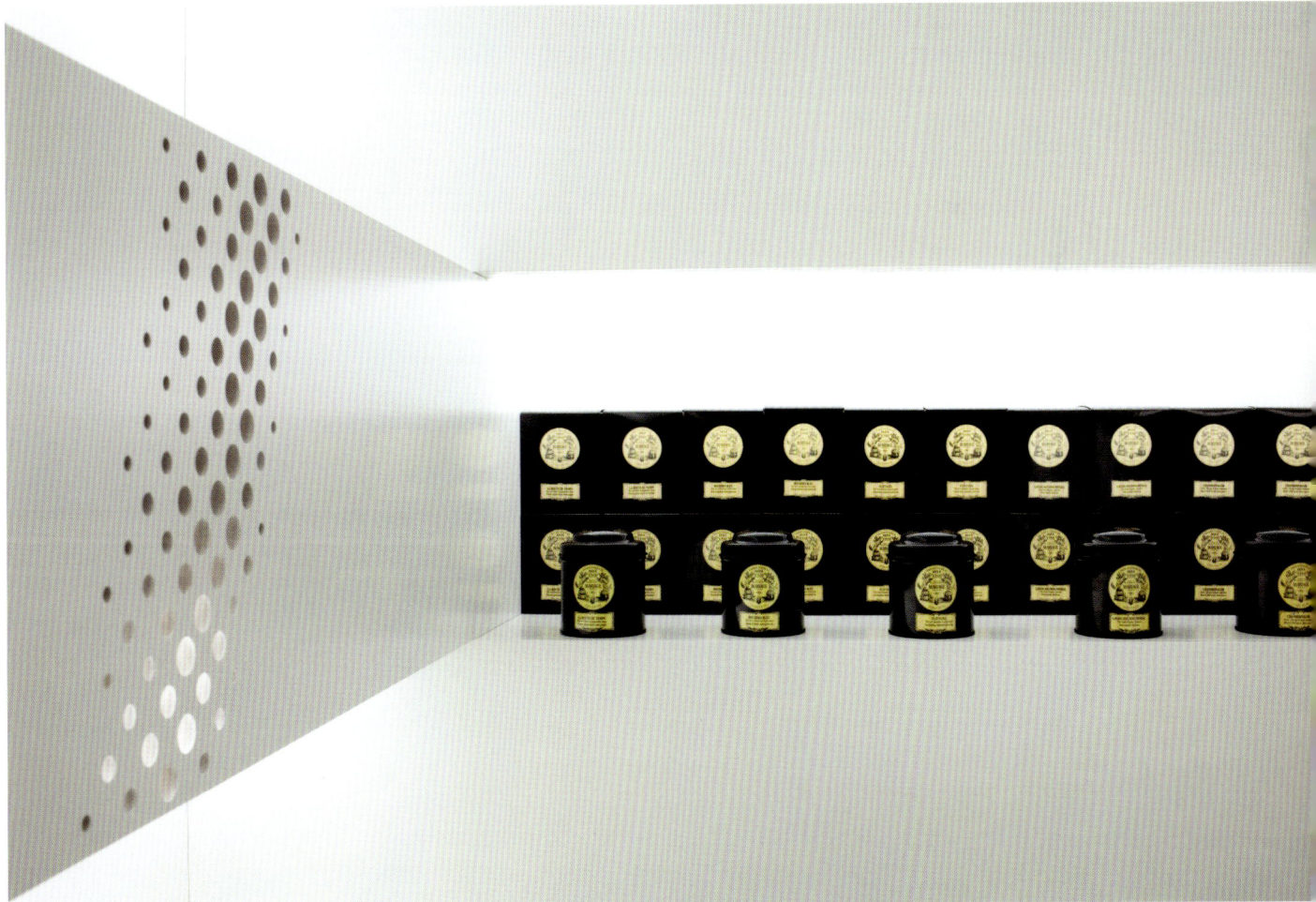

Designer: Gitta Gschwendtner
Client: Lik+Neon shop
Photography: Uli Schade
Location: London

Lik+Neon Shop

Design shop Lik+Neon have commissioned the designer Gitta Gschwendtner to redesign their eclectic and vibrant shop in London. Lik+Neon sell a unique selection of products including T-shirts, art magazine, interior objects, jewelry and art pieces. The display concept for the varied stock explores a juxtaposition of order and randomness, cleverly integrating bungee cord to create tidy grids showcasing the beautiful covers of the magazines, CDs and records, each one of them practically a piece of art in their own right.

In dynamic contrast, square display pegs jut from the walls in apparently random fashion, creating sculptural protrusion that function as hocks for prints, T-shirts and jewelry. The pegs create a pixilated effect continued by Gschwendtner's striking one-off ceiling installation—a lighting system devised from hundreds of plastic milk bottles, creating three glowing abstract clouds illuminating the white interior.

Lik+Neon是一家设计商店，销售一系列风格独特的产品，包括T恤衫、艺术类杂志、室内饰品、珠宝和艺术品。对于不同商品的展示理念，设计师寻求一种秩序感和随意感的平衡。橡皮筋经过精心的设计，形成整齐的网格，可以用于展示杂志、CD或者唱片的封面设计，因为它们自身就是一件件漂亮的艺术品。方形的展示挂钩似雕塑一般从墙面上伸出，也可以用于展示商品，这是设计中的动感元素。上百个塑料牛奶瓶被制成一组照明系统，看似发光的抽象云彩，点亮了白色的室内空间。

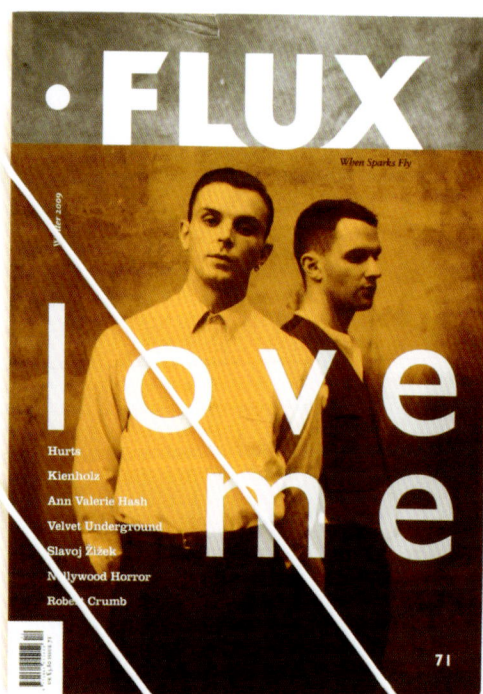

3 Legged Red Fox
Vulpes vulpes
Cockroach
Blatta orientalis
Wild Life of London
Flying Ant
Formica rufa
Crack eating Grey Squirrel
Sciurus carolinensis
Feral Pigeon/Rock Dove
Columba livia
Common/Brown Rat
Rattus norvegicus
Gärudio Studiäge

Designer: nendo
Client: 24 Issey Miyake shops
Photography: Daici Ano
Location: Tokyo

24 Issey Miyake

Based on the concept of the Japanese convenience store, 24 Issey Miyake shops combine inexpensive prices, a large variety of colors and frequent changes in product lineup. The Miyake team wanted a new design concept for the 24 Issey Miyake shop in Shibuya's Parco shopping complex, which includes a store that specially features Miyake's new Bilbao bag.

The Bilbao bag has no set form. Instead, it settles depending on how it is placed. To match the bag, we abandoned the standard hard, flat and smooth fixtures found in most shops, and created a set of variable-height fixtures made of thin steel rods that stand like a field of prairie grass in the shop, with a similar vague, undefined shape like the bag. Shelving and hanger rods are also made of steel rods, in the 7 mm diameter common to all of the 24 Issey Miyake shop interiors. Supported by 'points', rather than by surfaces or lines, the bags seem to waft in the air like flowers in a light breeze, creating the illusion of a field of flowers in the store.

24 Issey Miyake商店位于一家购物中心之中，以销售物美价廉、色彩多变的服饰产品为主。设计师以日式便利店为设计灵感，突显了Miyake的新产品Bilbao手袋。因为这款手袋无固定的形状，它会随着所承装的物品改变，所以为了强调产品的这一特点，设计师没有选用标准的陈列架，而是自创了一种以细钢条组成的错落有致的陈列设施。以这些"点"而不是面或者线支撑的手袋宛如微风中的鲜花一样，营造了一种花园的错觉。

Architect: Claudia Meier
Location: Meilen/Zurich, Switzerland
Client: Manuela Daluz
Photography: Claudia Meier

Boa Hairdressers Salon

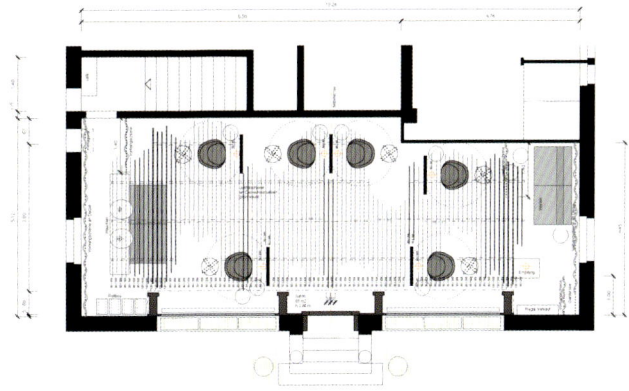

The new interior of 'Boa Hair' is giving the impression of entering an own world, a new identity of the hairdressers salon—the space has given a hair cut. The complete underside of the ceiling is covered with hanging white-transparent fibers. A slight movement circulates in the fibers when hairdryers blowing air through the space. The development and production of the material and elements took place together with Wasag AG Switzerland.

这家美发沙龙的新室内空间营造了一种美发沙龙的新模式，即空间本身好像也进行了一次头发修剪一样。天花板的下面全部覆盖了悬垂的白色透明的纤维。当发型师使用吹风机的时候，也会引起这些纤维的微小移动，带给客人一种非凡的空间体验。

Designer: Rafael de Cárdenas
Location: New York

Nike Sportswear's Stadium

New York based architect and interior designer Rafael de Cárdenas has designed Nike Sportswear's Stadium, which coincides with the summer celebration of the FIFA World Cup. De Cárdenas collaborated with Nike to create a space where visitors can experience product, design, and innovation, combined with the worlds of film, photography, art and music.

In the retail space, pegboard, a utilitarian commodity, is highly worked to elaborate effect and immerses patrons in a field often overlooked as a defining pattern of commerce. De Cárdenas often employs everyday materials in his conceptual design propositions.

The main event space is designed to perform multiple functions. Polygonal modular cells dispersed throughout move and change like players on a field under a directional, pitch-like pattern of linear fluorescents. The arrangement of the cells according to event creates micro-spaces within the larger context to enable installation art, games, or other purposes.

Says de Cárdenas: 'We began with a soccer field as a visual start and rotated it, revolved it and replicated it until we eventually came up with a unique pattern made of directional lines. We then applied this pattern to the floor to give it the dynamism of a sports/football field. The design allows for people to move the triangular shapes and customize their own experience at stadium.'

From architecture to the visual arts and event programming, Stadium NYC will continue to look to various artistic partners to deliver creative expressions of soccer throughout the summer and into fall 2010.

为了契合2010年盛夏南非世界杯的庆典活动，本项目设计师与Nike品牌合作，创造了一个参观者可以体验产品、设计和创意，并融合了电影、摄影、艺术和音乐的空间。在零售区，洞洞板这个实用的物件有力地营造了环境，其中还隐藏了赞助人的名字。主要的活动空间具有多种功能。多边形的模块分布在四周，就好像在球场上踢球的球员们一样，它们是为了实现装置艺术、游戏，或者其他目标而摆放的。地面上富有动感的线条图案是设计师以球场为灵感创作出来的，赋予参观者独特的体验。

Designer: Koji Yamanaka, Yuji Yamanaka, Asako Yamashita
Design Office: GENETO
Client: DIESEL JAPAN CO.,LTD.
Photography: Masato Kawano (Nacasa & Partners Inc.)

Power Plant

The designers recognized this space as a transmission base or power-generating place; a power station, filled with energetic atmosphere and conveying a message to the world. This 'power station' delivers DIESEL's cool message worldwide.

Composed of twenty-eight pinkish-red, wooden panels in zigzag lines, the installation dynamically changes the visitors' views and controls the path through the space to make visitors experience the shop environment.

As the panels are made of two plywood boards and are so heavy, the designers designed the zigzags as freestanding structures. Small screws on the panels are important aesthetically and practically. Rough wood surfaces look cool and are richly suggestive of DIESEL brand images.

POWER PLANT
DIESEL DENIM GALLERY AOYAMA

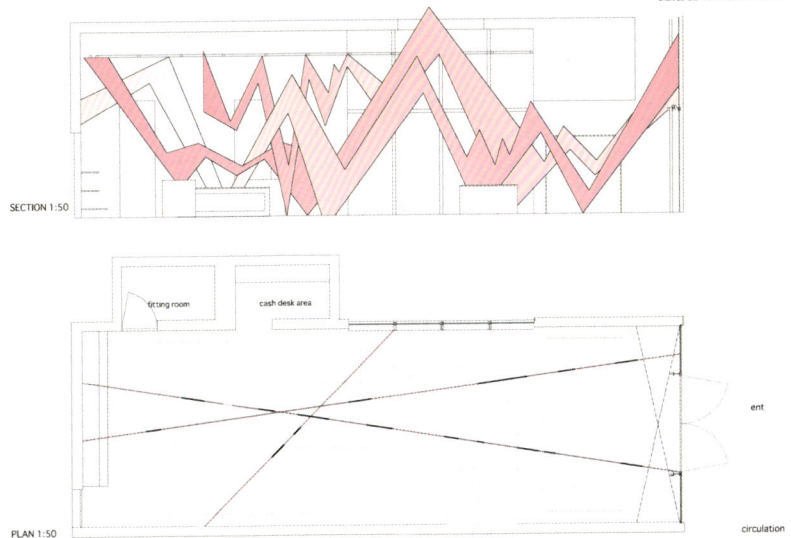

SECTION 1:50

PLAN 1:50

fitting room   cash desk area

ent

circulation

GENETO Architecture Office

设计师将这一空间看做是能量转换基地或者发电厂。这是一个动感十足的空间，DIESEL服饰品牌的内涵被传达得淋漓尽致。空间装置是由28跟桃红色的之字形镶板组成的，它们使参观者的视角发生改变，并控制了行进路线，参观者可以尽情享受空间体验。由于镶板是由两块胶合板组成的，非常重，所以设计师采纳了之字形的独立结构。镶板上的小螺钉既美观又有实际的功用。粗糙的木质表面看起来非常酷，这与DIESEL服饰的品牌形象完美吻合。

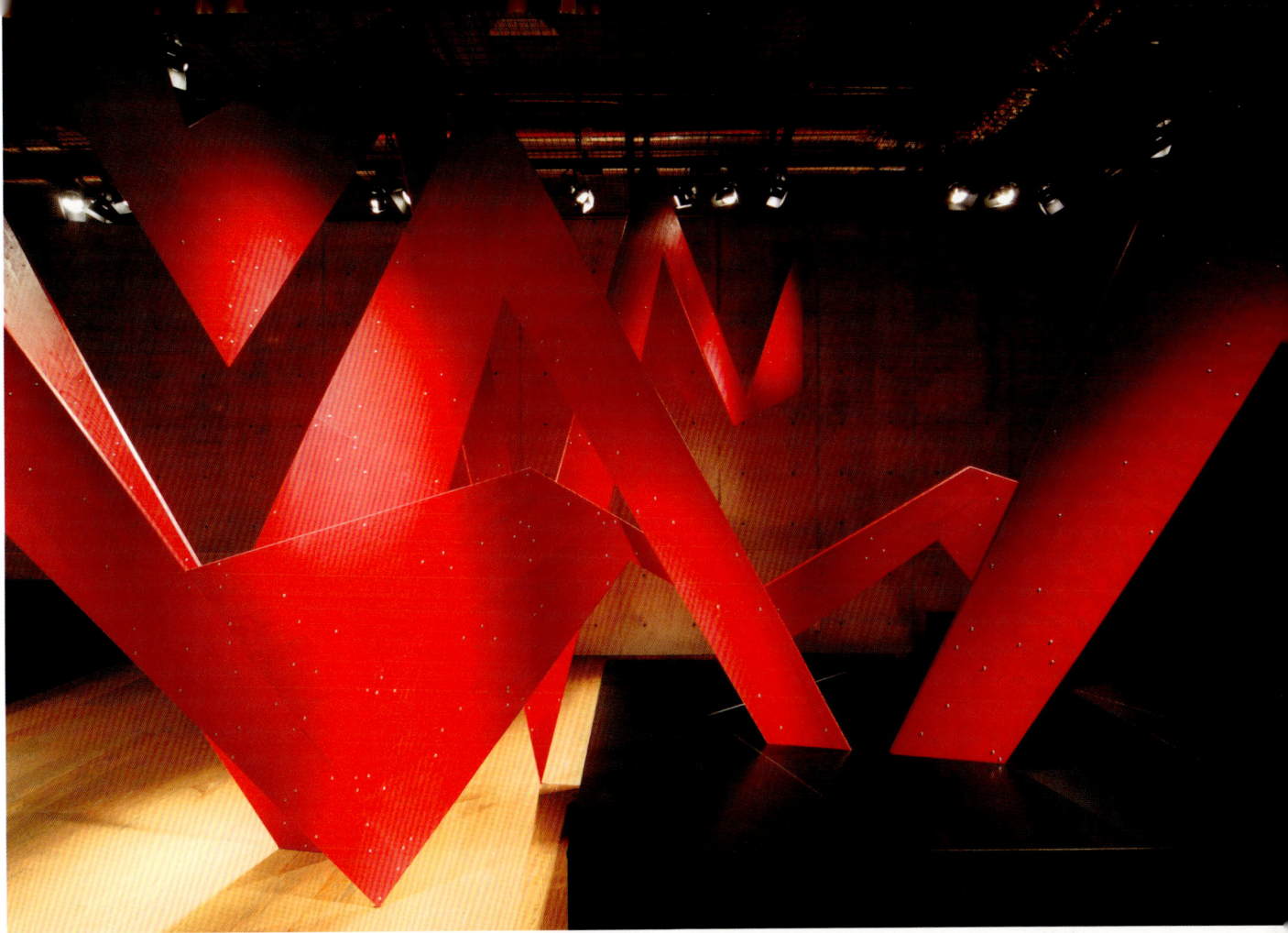
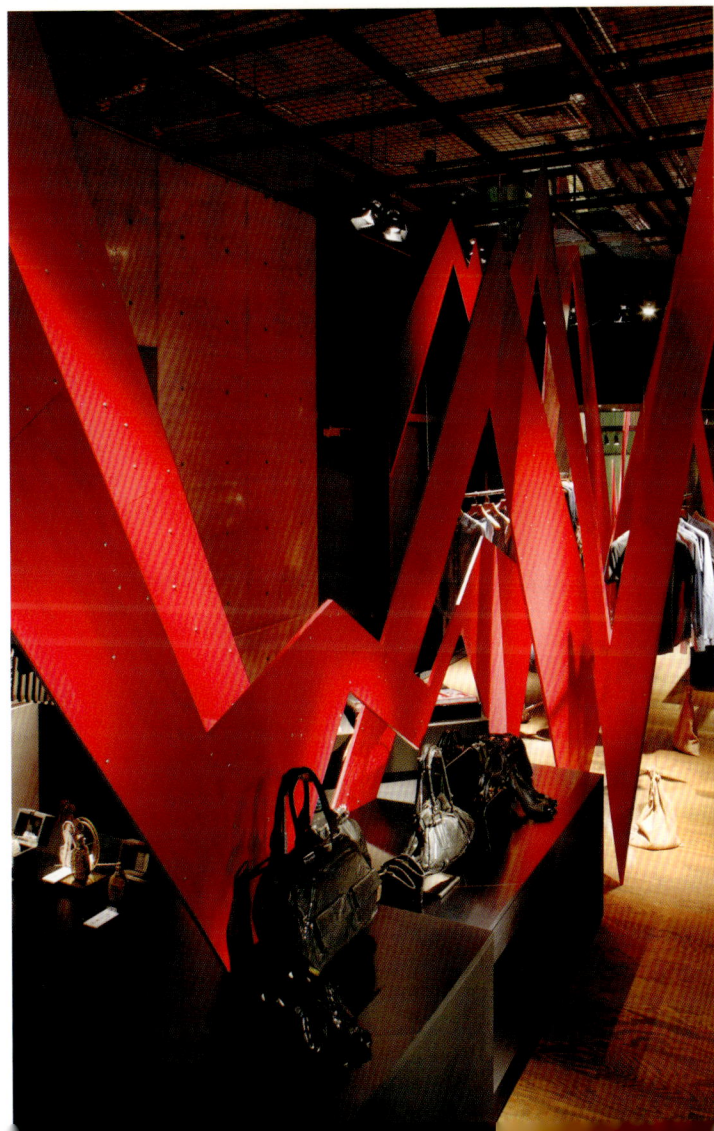
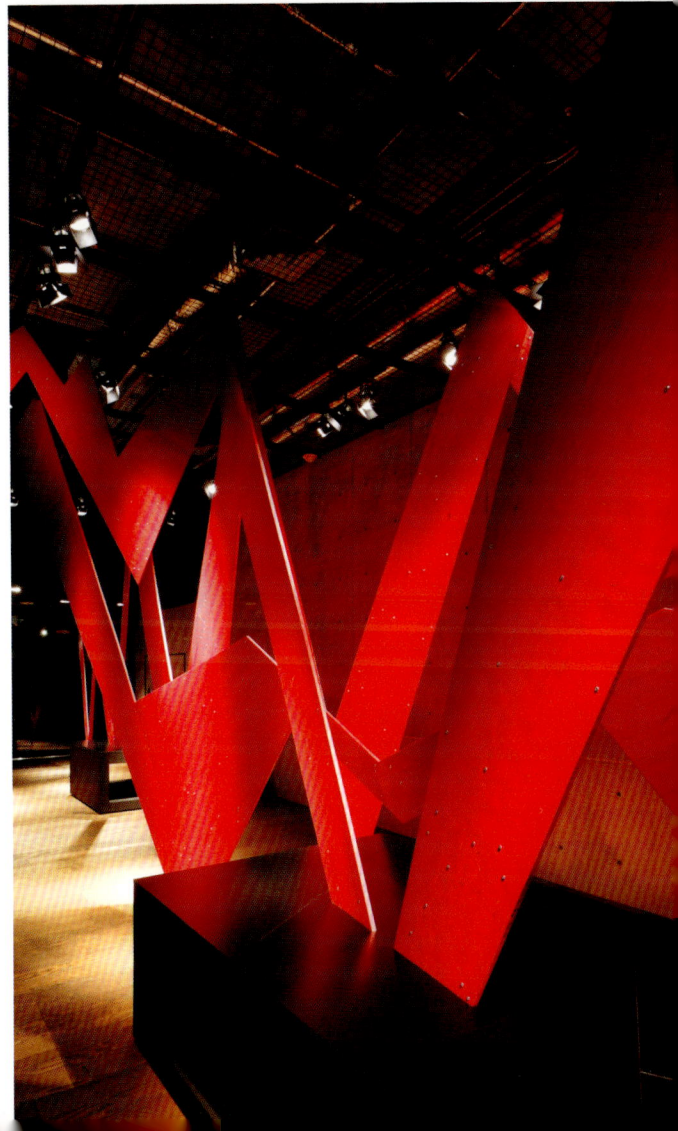

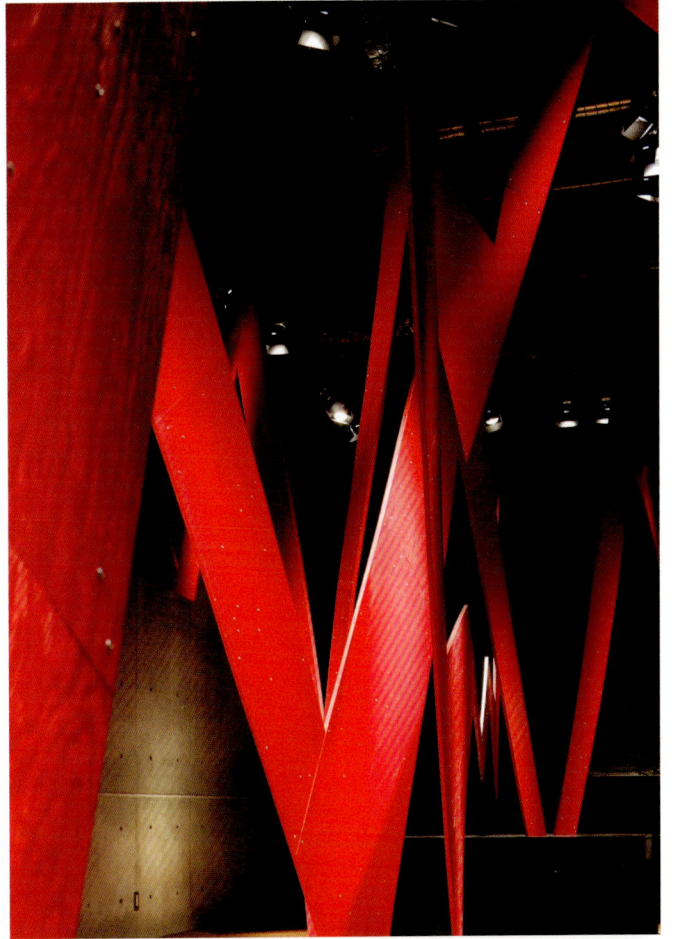

RESTAURANT

Designer: Morphogenesis
Project Team: Sonali Rastogi, Neelu Dhar, Manoj Kaundal,
Anika Mittal, Shilpa Puri
Location: 35, Defence Colony Market
Total area: 148 m²
Photography: Andre J Fanthome

United Coffee House

Box in a Box -the Concept of the Fine Dining Restaurant

Jaali wrapover on the exterior facade

Harshingaar Motif -the Main theme of the interiors

Harshingaar Pattern engraved on the False Ceiling and the AC Ducts

Bamboo Lamp Installation in the Staircase

Harshingaar cut in wood and configured on the steel railing

Harshingaar Motif blended on the interior jaali

Harshingaar pattern replicated on the furniture

Sequel, by United Coffee House is located in a busy market square in the heart of the one of the up market colonies of New Delhi. United Coffee House is one of the old and renowned restaurants of Delhi that has deep-rooted memories for most Delhiites who frequented the restaurant over decades. It is a new venture by the people who own United Coffee House; almost a sequel to the existing legend in Connaught Place.

The market square is extremely busy and an identity for the restaurant was a part of the design brief—both for the passers-by and for the restaurant customers. The Defence Colony market is a 50s-60s Indian modern format of High Street shopping, catering to the needs of the local residents. Hence, the shop sizes were narrow—10 ft wide and initially, were only for local trade. Over a period of time, the market was over taken by the pressures of Modernization and hence, most of the shops there are a patch-fit story. The existing situation is of clutter and all that one sees while walking around is a mess of wires, signages, services and air-conditioning. There is no attention that is paid to the exterior and it is almost impossible to recognize one single building by itself. Morphogenesis saw an opportunity to do an insertion to lay the seeds of an urban renewal, and as a challenge to establish an identity for Sequel through this opportunity. Although initially commissioned as an interior design project by a client who luckily owned almost this entire building, we saw the project as a possibility to create an identity within this urban spread and move away from the disorder. Hence, the approach was outward-in. Owing to the narrow depth of the shops which were dark, dingy and damp, the design was conceptualized with interplay of light and shadow as an architectural feature by means of a 'jaali'.

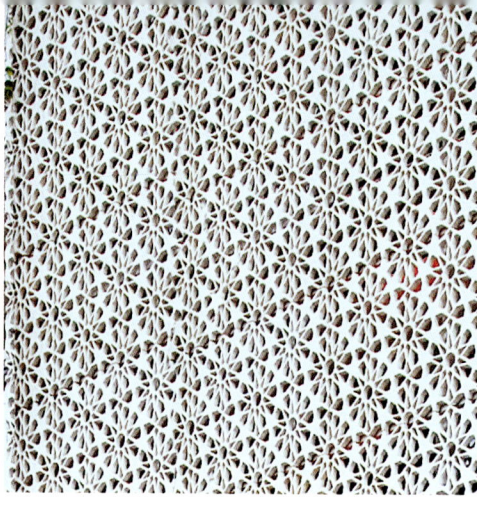

The brief also required a sense of privacy and pampering—transporting the food aficionado away from the clutter and noise of the street to a calm, serene, and sophisticated Zen-like ambience. The interior theme was derived from two main notions; one, the idea of a [Box]² —A Box within a Box. This would help to provide individuality and identity within the cluttered visual of the narrow side-by-side high street buildings. Secondly, the use of a modern version of the 'Harshingaar' motif signifies the contemporary, yet traditional nature of the space. To blend in the contemporary nature and function of the new fine dining/lounge with the traditional paradigm of a family restaurant that United Coffee House stands for, this time-honored motif of 'Harshingaar' has been used symbolically as an integral part of the design theme. (The Harshingaar is also called Night Jasmine or Tree of Sorrow)

Fitted within an extremely linear plot in the market, it was a huge challenge to accommodate a fine dining restaurant and lounge within the tight setting of the site. The linearity of the site was very demanding, and it was a task to accommodate

the number of people that could be eventually seated within the site. Since there is so much new insertion going around the market, the intent was to exploit this opportunity to create a magic that was constantly changing and attempt to ensure that the restaurant does not become dated.

The restaurant space is conceived as a discreet high design space, strung along an almost sculpture like staircase moving over three floors. The staircase epitomizes the entrance and creates an impact at the heart of the building. The 'Harshingaar' is cut in wood and used in multiple configurations as a part of the staircase railing. A bamboo lamp installation over this triple height space created by the staircase visually creates an impact by means of its exceptional sculptural identity. A traditional skin— 'jaali'— is wrapped all over the exterior facade using lights and voids as a pattern-making tool. The pattern is abstracted from the 'Harshingaar' and is used to create dramatic and dynamic light changes through the day and over seasons. At night, the light changes can be programmed for festivities.

LONGITUDINAL SECTION A
SCALE 1 : 60

LONGITUDINAL SECTION B
SCALE 1 : 60

The current thoughts prevalent within the practice were to reinterpret and revive traditional Indian craft and tradition and give it a lifeline versus its current status of dying a slow, neglected death. Further, the idea was to modernize craft and hence, abstraction of the 'Harshingaar' was done in a multitude of ways and its presence can be seen all through the interiors of the project; on the walls, the furniture, the staircase, the ceiling, and lights etc. The craft and the abstraction of the motif of 'Harshingaar' was designed and customized together with India Chic (Mike and Preeti Knowles).

The pattern is used in the form of both reliefs and perforations that abstract the theme in an extremely contemporary manner. Once inside the building, a wooden 'jaali' with the 'Harshingaar' pattern demarcates the seating areas. The pattern also finds place on the furniture—the back of the chairs, the tabletops, and even the appliqué work on the walls of the lounge areas. The false ceiling and the AC ducts are also clad in wood with the 'Harshingaar' motif cut out to bring in light into the restaurant space, almost making it glow in light. It is this pattern and the vocabulary of the furniture that connects all the three floors of the restaurant—the booth areas, the communal table areas and the Lounge area. A limited palette of colors and materials is used—beige, wooden floors and red is used to highlight the features.

An urban seed has been used as an insertion into the existing traditional Indian modernist market streetscape. Sequel is an attempt to break away from the typical clutter, through a simplistic alternative rooted in the evolution of the 'jaali'. The intent is to encourage design and materiality that is deh rooted within the region. What is unique to the high design nature of the restaurant is that the entire design was treated as a high design handcrafted product. The final outcome is exclusive, memorable and will probably age well; from engaging and stimulating the old customers, to welcoming the new, the Sequel is more than just an experience

这间餐厅坐落在新德里闹市区的高档消费中心，是联合咖啡屋的升级改造项目，其前身是新德里最古老的餐厅之一，几十年来承载着新德里人的美好回忆。这间餐厅周围的城市环境非常嘈杂，而且原有室内空间狭长昏暗，因此打造一间远离城市喧嚣、迎合城市更新主题的新餐厅对设计师们来说，确实是一个挑战。设计师依据两条设计思路做出了最终的方案：一是，盒子中的盒子，呼应周围的城市环境，同时体现独立性；二是，借用夜来香的图案，体现餐厅的传统和现代。餐厅的三层空间以一个雕塑般的楼梯相连，木刻的和镂空的夜来香图案被用在不同的构造中，如楼梯栏杆、墙壁、餐椅和隔断等。这间餐厅的设计本质体现出一种手工艺品的特性，以精致和回忆吸引着新老食客。

Designer: Stylt
Photography: Erik Nissen Johansen
Location: Stockholm

Griffins' Steakhouse

Once again, the storytellers and designers at Stylt have created a successful restaurant concept that puzzles the industry. What's the secret with these Scandinavian wizzards? Less than a month after it's opening, the latest Stylt offering, Griffins' Steakhouse Extraordinaire in Stockholm, has already gained rave reviews and international recognition. And the guests just keep a-coming.

As a very intense year draws to it's end, the highlight of 'styltism' appears to be the Griffins' Steakhouse Extraordinaire in Stockholm, located in the high-profiled Waterfront Buildings. 'Stylt has done a fantastic job and the guests are lyrical about the interior', says Karl Ljung, CEO of Griffins', where the fictional host couple's penchant for mysticism and alchemy is reflected in an environment unlike any other.

If Griffins' had been located in Chicago or New York, it could very well be seen as a historic landmark. The culinary quality goes hand in hand with an interior design that is both innovative and timeless. The cozy atmosphere is a warm contrast to the Waterfront Buildings otherwise austere design. The Griffins' are an interesting fictional host couple made alive in almost scenographic furnishings with science,

romance and aesthetics as main ingredients. The restaurant's classic American-style is spiced up with artifacts and travel memorabilia in an atmosphere combining the good life and its unexplainable sides. That is what it looks like in the home of devoted chemist Griffin and his glamorous wife.

'We have strived for a home-feeling but with a distinct style and dignity,' explains Erik Nissen Johansen, responsible for both concept and execution on behalf of the owner Stureplansgruppen. Griffins' Steakhouse Extraordinaire is Stylts third strike in Stockholm in a short time. While most major players in the industry still settle for the standard solution of replication and stereotype design, Stylt keeps steering in the opposite direction. Erik Nissen Johansen explains how this strategy allows Stylt's clients to stand out in an extremely competitive market: 'It is very simple. Hotels and restaurants that incorporate personality differentiate themselves from the existing offerings, and therefore are able to maximize guest satisfaction and increase the return rate. We've proven this to be a financially sound strategy. Replication may save money in the short-term, but it may alienate those guests looking for a more unique experience.'

擅长"讲故事"的设计公司Stylt在斯德哥尔摩引人注目的港口地区又成功打造了一家餐厅。在开业不到一个月的时间内，餐厅获得的好评如潮，食客们络绎不绝。餐厅的主管说："Stylt的设计精彩极了，客人们对室内环境赞不绝口。"如果这家餐厅位于美国的芝加哥或者纽约，那么它极有可能被视为一处历史地标。其提供的餐品与室内设计都富有创意并经久不衰。室内布景将科学、浪漫和美感相融合，以手工艺品和旅行纪念品营造出了传统的美式风格。设计师说道："这里有一种家的感觉，又具有独特的风格。客人们喜欢再次光顾具有自身特色的酒店和餐馆。复制相同的设计虽然可以获得短期利益，但客人们追求的是更独特的空间体验。"

Designer: design spirits co., ltd.\Yuhkichi Kawai
Client: Caesars Palace
Location: Las Vegas
Total area: 297 m²

Beijing Noodle No.9

This modern Chinese restaurant, Beijing Noodle No. 9, is located within the huge casino hotel holding more than 3,300 rooms in Las Vegas. The restaurant is adjacent to the casino; consequently, the excitement, gaming machine sounds, and neon lights are naturally overflowed into the space.

In general, a space usually consists of various interior elements, materials, series of products, and patterns placed appropriately. Taking advantage of no support columns in the site, the Beijing Noodle No. 9 utilizes the only one pattern throughout the space to achieve the minimal and visually calm atmosphere set in the heart of a surrealistic environment.

Because the restaurant is situated next to the casino, an isolated space cooling the excited customers down is not ideal. Therefore, the boundary between the restaurant and the casino should not be clearly defined but subtly celebrated. In other words, restaurant should become not only a respite but also an extension of the casino.

The restaurant incorporates the subtle boundary bringing the fading casino sounds into the space to create the inviting welcome to customers. Since the open design facade allows the sounds in and out, the one primary arabesque pattern is applied to the entire space to produce visual calmness.

As customers walk into the space, they will undergo the visually and physically unblocked entry experience, and are comfortably led past the sparkling aquarium tanks to the deep part of the restaurant. The gorgeous double-wall design features the elegant arabesque patterned layers, consisting of one woodland-patterned steel decorative surface above a painted similarly patterned solid back, with glossy finish. This design creates the arabesque shadow effect, and the wall continuously extends to be the ceiling, which generates a soft cocoon-like interior experience. By placing LED indirect lights between the layers, the arabesque patterned shadows are appeared on the tables, floor, and kitchen appliances to accomplish a cohesive design throughout the space.

The overall dining experience, supported by brightly lit ambient light, envelopes customers in a sensuous embrace. Dreamy lighting contrast projecting the wonder of deep forest, and secrets of the ocean conquers the sounds from the casino to create a visual serenity.

本项目设计是一个现代中餐厅，位于拉斯维加斯一座拥有3300多间客房的酒店内。设计师利用了场地中无支撑立柱这一优势只在空间中使用了一种图案元素，就营造出了超现实的视觉氛围。当食客们步入餐厅的入口区时，他们会同时获得视觉上和身体上的双重感受。闪闪发光的水族箱为他们指引了通向餐厅深处的通道。华丽的室内墙面和天花由表层的钢制表面和底层的石膏板组成，而阿拉伯风格的花纹图案是这一设计的点睛之笔。表层与底层之间的间接LED光源使花纹图案的光影效果映衬在餐桌、地板和厨房设备上，所有这些元素共同为食客们营造了一种被森林仙境环抱的感觉。

Designer: UXUS
Client: Selzim Restaurant Group
Location: Sacramento California
Total area: 710 m^2

Ella Dining Room & Bar

ELLA RESTAURANT AND BAR

12TH & K STREET, SACRAMENTO, USA

GROUND FLOOR PLAN

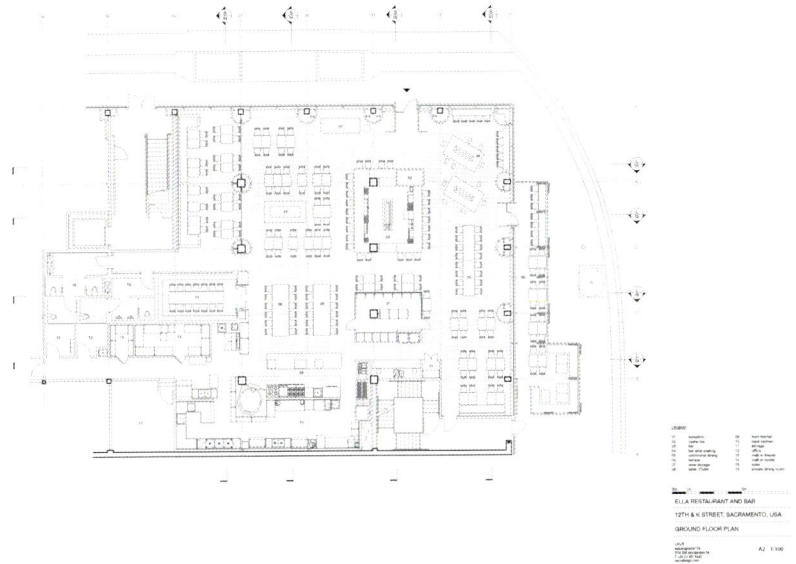

Ella Dining Room & Bar serves 'Modern American Bistro' cuisine. The owners wanted the restaurant to become 'Sacramento's living room', an urban oasis where lawmakers and other diners can go and unwind after a long day's work.

The design objective for Ella is: create a brand that embodies the principles of 'Rustic Luxury', and that celebrates an elegant, relaxed contemporary lifestyle. 'Rustic luxury' is a synonym for purity, the essential beauty and goodness contained in simple things. It is about the pleasure and sensuality of real materials, and about the inherent comfort of a natural, effortless style. 'Rustic luxury' is not a simplistic reduction. It is the magical crystallization of two apparent opposites, simplicity and complexity. Rustic Luxury, as defined by Ella, offers its guests an experience that combines the simple and natural pleasures of dining at a dear friend's home, at their table d'hôte or host's table.

The 'Host's Table' is a French tradition of eating in the kitchen while the chef is preparing dinner. It is a very welcoming and intimate experience, usually reserved for honored guests and close friends. The owners of Ella wanted to create that level of intimacy at their restaurant. UXUS cleverly opened the kitchen to the dining area with 2 large communal tables forming the table d'hôte area. Diners can experience the thrill of watching the chefs at work, and taste the results of their culinary efforts.

All of these elements come together at Ella Dining Room & Bar to form an intimate and convivial dining experience, the embodiment of 'Rustic Luxury' right at the heart of California's State Capital.

Ella餐厅专门提供具有现代美式风情的餐品，这是一处都市绿洲，是食客们在一天工作后可以放松身心的地方。因此项目的设计目标是：营造一个体现"乡村奢华"的品牌，体现优雅、休闲的当代生活。"乡村奢华"代表的是洁净、纯美和简约的风格，是一种由自然和原始材料带来的愉悦舒适的氛围。食客们就像在朋友们家里就餐一样，体验着简单的快乐。在法国，人们有时会在厨房招待尊贵的客人和亲密的朋友，这是很私密的体验。设计师将这种传统移植过来，开放了厨房，使食客们可以一边欣赏烹饪的过程，一边品尝厨师的手艺。

Designer: concrete architectural associates
Client: Al Jaber
Photography: Richard Thorn
Total area: 1430 m^2

Pearls & Caviar

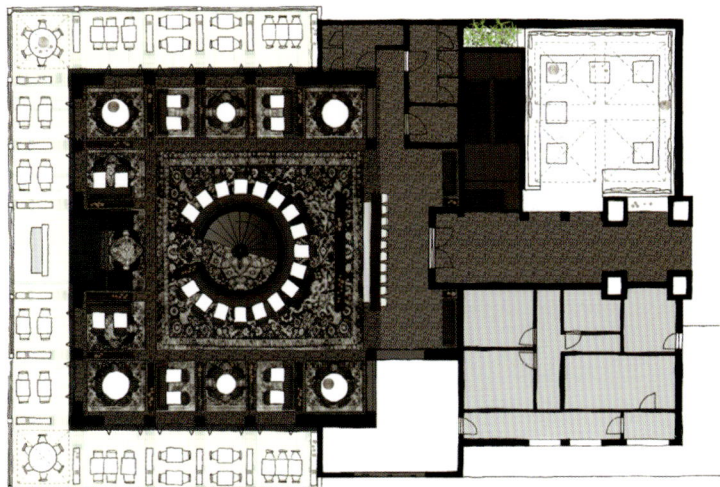

'Pearls & Caviar' represents the new Arabian lifestyle, a luxury fusion of east and west, black and white, occident and orient, light and shadow, the extrovert and the introvert, the intimacy and the view.

The basic idea of the design of the restaurant has been to create an abstraction of the commonly used oriental forms and materials without loosing their richness. To achieve this the amount and the richness of the oriental patterns and forms which are found in traditional oriental spaces are kept but without color. All colors are replaced by either shades of black or shades of white, both in combination with silver. Therefore the restaurant is divided in two parts, which also gives its name.

　"Pearls & Caviar" 餐厅是对阿拉伯风格的全新演绎。设计师从丰富的东方设计形式和材料中提炼设计灵感，去除丰富的东方色彩元素，代之以体现现代感的黑色、白色和银色，将东与西、光与影、内敛与开放的对立感完美融合。

'Pearls':

The 'pearl' part is represented in the terrace on level 1. The terrace is divided in 3 subspaces: the main terrace and two private terraces, which can be used separately. The pearl areas are extrovert spaces composed completely in shades of white and silver.

The flooring is a big oriental carpet made of glass mosaic tiles in the white-silver color range. A circular bar made of polished stainless steel with a bar top of white marble is placed around the tower of the main terrace. White

upholstered, deep lounge benches are placed along all borders of the single terraces and white sails cover each of them as sun protection and light diffuser.

"Pearl"是露台区域的设计主题词。地面呈现的是由玻璃马赛克瓷砖拼铺而成的大型东方地毯图案，白色软包座椅环绕在四周，其上的白色遮篷可以提供阴凉和防止光线漫射。

'Caviar':

The 'caviar restaurant' is situated on level 0. It is an introvert and intimate space in the inside of the building.

The restaurant is divided in little intimate 'chambers' on the rim of the carpet separated by ball chain curtains and side tables made of wengé. On top of them Persian accessories are placed which function as light diffusers. Around the central tower a circular bench is placed upholstered in black leather. The back parts of the benches stretch over the full height of the wall of the tower. And a straight bar in stainless steel with black marble bar top.

"Caviar"餐厅是整体设计中的私密区域，由珠帘分隔成一个个包间。位于中心位置的黑色软包环形座椅的靠背一直沿墙而上，而放置于边桌上的波斯饰品还起到了漫射光线的作用。

Designer: Yuhkichi Kawai, Akiko Orii
Design Office: design spirits co.,ltd.
Photography: Zainudin & Toshihide Kajiwara
Location: Malaysia
Built-up area: 652 m^2

Rootz

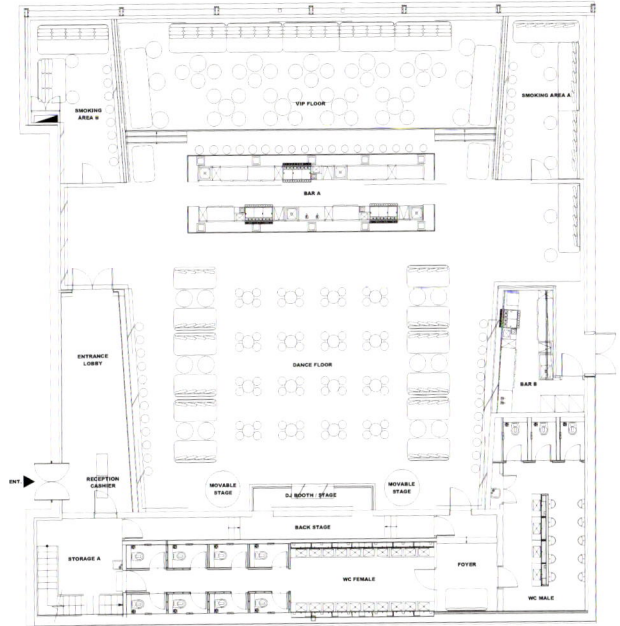

This club lounge, Rootz, was planned in Kuala Lumpur Malaysia, the roof parking space of the existing shopping center which greets the 20th anniversary. Fortunately, the designer was able to plan it by himself about a club, a theater, and a restaurant, the place of the courtyard or each area because he was able to participate from a master plan. It has taken 1 year and six month from a plan to the completion to reinforce structure.

The designer said: 'there was a demand to regain the youth who left it by I resembled the roof of the shopping center newly, and making a club lounge. I thought vaguely if I said that I danced, there was the certain figure from old days whether in the East or in the West to dance together in the grand hall of the palace. Also I seemed to match this thought in the times of the gorgeous trend in now days. The plan was adopted soon, so I negotiated patiently with the palaces of various countries and took permission in the copyright from the Russian palace at long last and went out with a photographer from Japan and had it take a photograph. I edit it and print it on form to fit a new skeleton as wall paper. Furthermore, I printed it on an organza and I hung it in front of the wall paper to make double layers and represent a ghost phenomenon.'

'I wanted to settle it for palace with rigid materials, but I didn't. I considered that it should be redecorated easily like repaper wall paper from the business face which is said to be short cycle club. I edited it not restoring the palace of the Middle Ages precisely and not to become artificial so that dimensions only encountered accurate. Then, I thought to create a bright illuminated club against the dark black club. The reason why I thought is that I feel in doubt and feel in corner cutting against the club which has dark illumination and paints both wall and the ceiling black to not realize. In the case of this project, the first idea and the negotiations of the copyright were going to be main of my work, and it was rare that I did so-called design. Even if it is a bright club lounge, this club is accepted to youth and record explosive sales. Drink is sold out, and 200 people whom it was not able to take inside waited in a courtyard, and some of them returned without can wait. I feel so relieved to hear that.'

设计师将位于马来西亚一个购物中心屋顶上的停车空间翻新为一间酒廊。因为参与了总体规划，所以他在会所、剧院、餐厅、庭院和每个区域的设计上拥有自由度。当谈论到跳舞，设计师的脑海中会出现在古时宫殿大厅中翩翩起舞的景象，这就是他设计本项目时的灵感之源。在得到了位于俄罗斯的一家宫殿的许可后，来自日本的摄影师拍摄了照片。设计师对照照片进行了编辑，制作成墙纸，并将照片印刷到硬纱上，悬挂在墙纸前面，营造出双重效果。不同于其他昏暗的会所空间，本项目光彩熠熠，并赢得了年轻人的青睐。

Designer: Rockwell Group
Photography: Eric Laignel \ Rockwell Group
Location: Dubai
Total area: 1068 m²

Nobu Restaurant

Nobu Dubai was an evolution of many of the concepts Rockwell Group developed for the flagship Nobu Fifty Seven in New York, such as the emphasis on craftsmanship, natural materials and storytelling. The textures and materials in this particular location were chosen to reflect the finely crafted cuisine and Nobu's roots in the Japanese countryside, while also celebrating the Dubai beachfront context.

David Rockwell, Founder and CEO of Rockwell Group, explained: 'The context, landscape and history of this new restaurant brought about all sorts of new and exciting challenges. We had to think about its location not only in the Middle East, but also in Dubai as the epicenter of an ever-growing and flourishing environment for building, not to mention being more specifically in the larger-than-life Atlantis resort on Palm Jumeirah.'

To anchor the restaurant in its location, Rockwell Group surrounded the restaurant walls and ceilings with large hand-woven curvilinear abaca panels to evoke an environment submerged under an ocean wave, and added accents of traditional Middle Eastern vernacular architecture such as hand-wrought iron columns of flowers, leaves and buds.

该项目是Rockwell Group继Nobu纽约旗舰店后开发的众多设计理念的一种演化，设计师强调了手工艺、自然材料和空间的叙事性。精心挑选的肌理和材料既反映了日式餐饮的精致，又强调了迪拜海滨风光的特质。餐厅所在地域及城市的文脉、景观和历史为设计师带来了全新的挑战。为了回应这一点，设计师以大型手工编织的弧形麻蕉板覆盖餐厅的墙面和天花，营造一种被海浪围绕的感觉。此外，设计师还重点强调了传统的中东建筑元素，如手工铸造的花朵、树叶和花蕊的图案等。

Designer: design spirits co., ltd.\ Yuhkichi Kawai
Client: AC2 International pte. ltd.
Location: Singapore
Total area: 502.75 m²

Nautilus Project

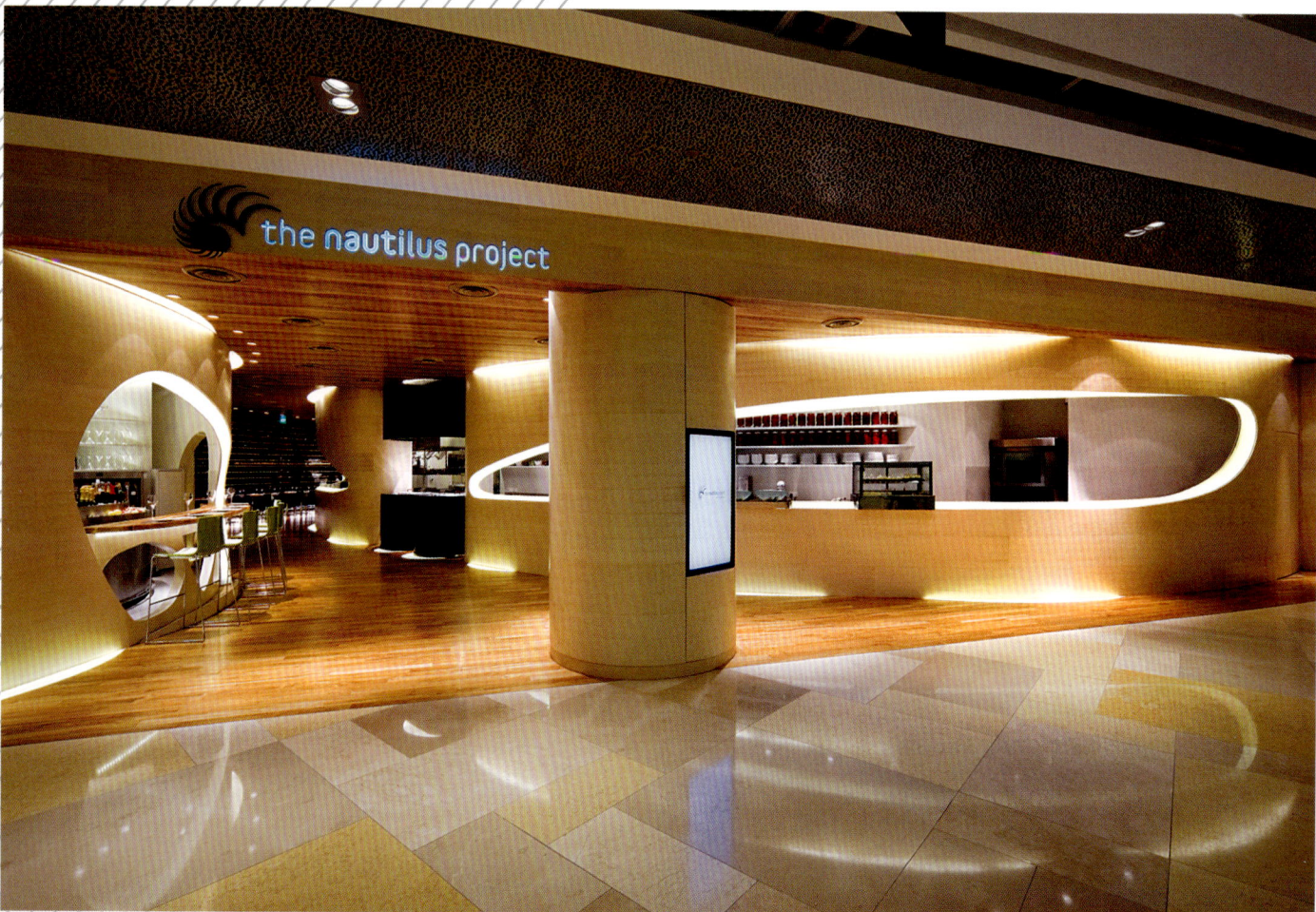

The Nautilus Project is located on the fourth floor of the ION shopping center, where opened recently on the Orchard Road, Singapore. This floor is not only the restaurant floor, but also there are shops as well. This floor is a composite both sales and drinks and eats. The beginning of the plan, the food consultant prepared a certain concept, chef, designer, and location, so the designer had a difficulty in leading his ideal interior design.

The owner is the president of the cargo company and, is a surprisingly beautiful woman boss, so the designer decided to reflect her sophisticate, elegant, and tender characteristic to this restaurant project. The chef was done recruiting of by owner oneself from New Zealand, and the chef was popular among New Zealand like a celebrity. The designer was able to never eat his dish until the completion of the project because he was very busy. Cooking is the leading role. The designer predicted that visitor unit price would rise in meetings, so he stopped making a facade and he designed the entrance where a common use passage led to restaurant to make visitor enter easily. The skeleton of the restaurant had curve, so he made the others curve to let it be familiar. The wood material was requested by owner. The designer made effort that it does not look futuristically because of its curve. It is a design lasting a long time not a style that the designer aims.

该项目位于新加坡一家刚开业的购物中心的四层。这一层容纳了餐饮、购物等功能。顾问公司在项目的规划阶段就选定了设计概念、厨师、设计师和地点，因此设计师不能完全按照自己的理念来设计。餐厅的业主是一位漂亮的女士，因此设计师决定将她的优

雅和温婉体现在项目中。入口区域的设计使食客们可以很容易地步入餐厅中，这是因为没有设置立面。餐厅的主体框架是弧形的，其他木质弧形元素的运用与主体框架相互呼应。设计师希望这种造型不会给人一种太未来主义的感觉。

Architect: Office dA
Project Design: Nader Tehrani, Monica Ponce de Leon
Photography: John Horner
Total area: 446 m²

BANQ

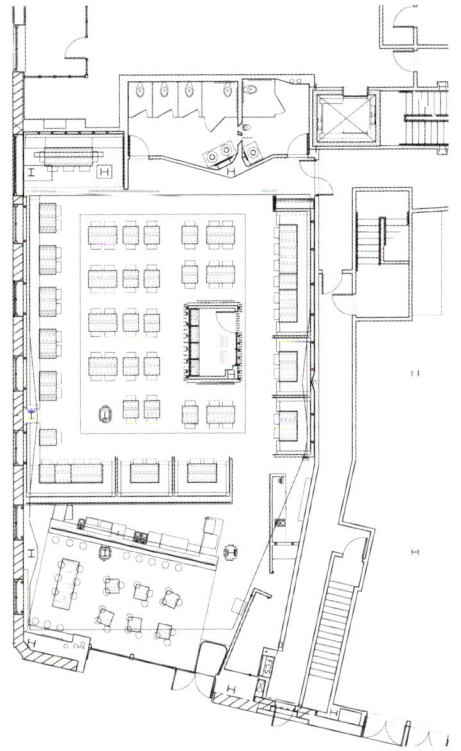

Located in the old Penny Savings Bank, Banq is a new restaurant located at the base of the old banking hall. Divided into two segments, the front area on Washington Street is programmed as a bar, while the larger hall behind serves as the dining area. The design of the space, however, is conceptualized around another division, on the z axis, between the ceiling and the ground. If the ground needs to remain flexible as a result of fluctuating activities of the restaurant space—two seaters, fours, and sixes, among a range of other organizations related to parties and other events— then, the ceiling contains fixed programs that are part of the building's infrastructure —the structure, drainage, mechanical equipment, sprinkler system, lighting, and other the acoustic systems.

To that end, we have developed a striated wood-slatted system that conceals the view of the mechanical, plumbing, and lighting systems on the longitudinal axis, while offering a virtual canopy under which to dine. The geometry of the wood slats conform to each equipment above, but are also radiused in order to smoothen the relationship between other adjoining equipment, creating a seamless landscape. The columns and the wine storage, in the middle of the hall, serve to uphold the fiction, and appear to be suspended from the ceiling. If the longitudinal axis emphasizes the seamless surface, then lateral views offer striated glimpses into the service space above, and demystify the illusion. To underline this strategy, certain areas of the ceiling 'drip' and 'slump', acknowledging the location of to place exit signs, lighting features, and other details.

147

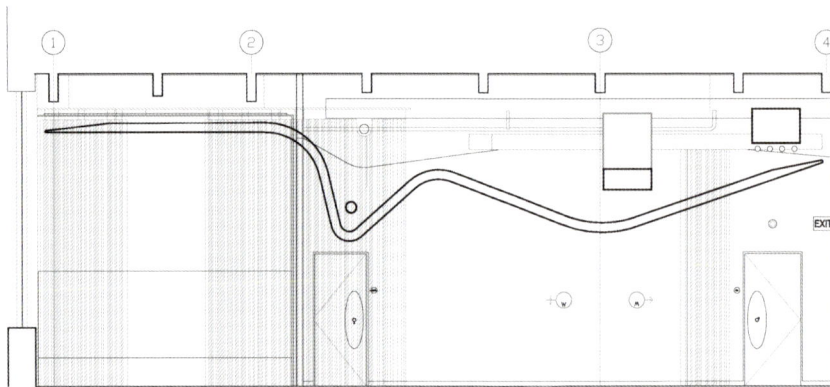

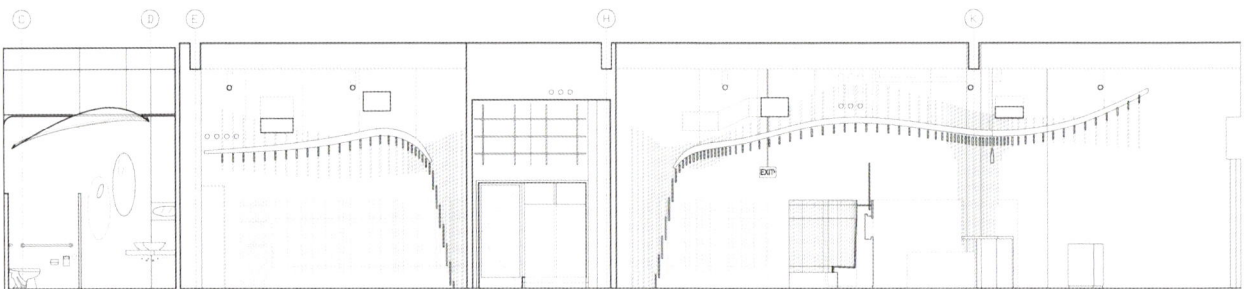

Below the ceiling, the functional aspects of a dining space are fabricated with warm woods and relaminated bamboo amplifying the striping affect already at play throughout the space. Striations of the ground, the furnishings, and the ceiling all conspire to create a total effect, embedding the diners into the grain of the restaurant. Acknowledging the historical setting of the building, the ceiling hovers away from all interior walls and instead finds its support in suspension from above. Nearly running the entire width of the space, each rib of the undulated ceiling is made from unique pieces of three-quarter-inch birch plywood adhered together in a scenario that likens to a puzzle; only one possible location for each unit, formulating the continuous member. These continuous members are fastened to the main structural ribs running perpendicular to the lattice, tracing both the overall ceiling topography and the steel supports of the base building. Spacing between the visible ceiling ribs is variable; compressing and releasing to maintain visual densities of the overall surface as seen from different angles.

Banq餐厅位于古老的Penny Savings银行建筑内，因此设计师获得了借用原有的空间感和功能性的优势开发新设计的机会，同时，如何使重新布置的就餐区域具有最大的灵活性，并克服旧建筑天棚上大量的机械、结构和声学问题，也对设计师提出了挑战。餐厅由两部分组成，靠近华盛顿大街的是一个酒吧，其后是较大的就餐区域。以一个条纹木板结构覆盖天棚是这一设计的解决方案，它既隐藏了原有设备和立柱，还为就餐者提供了精美的室内景观。

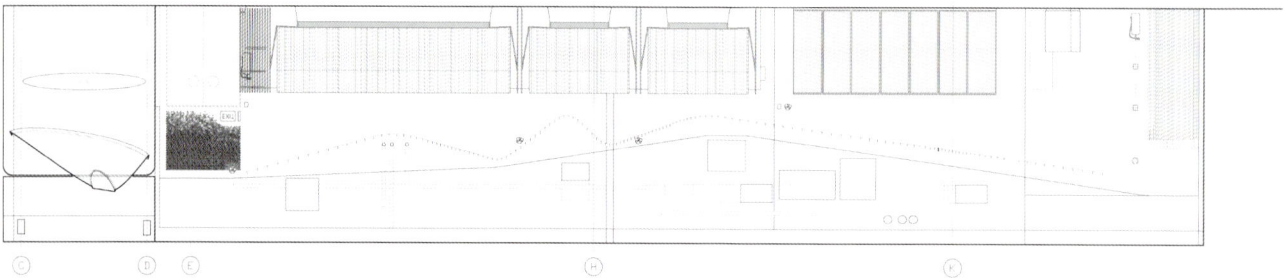

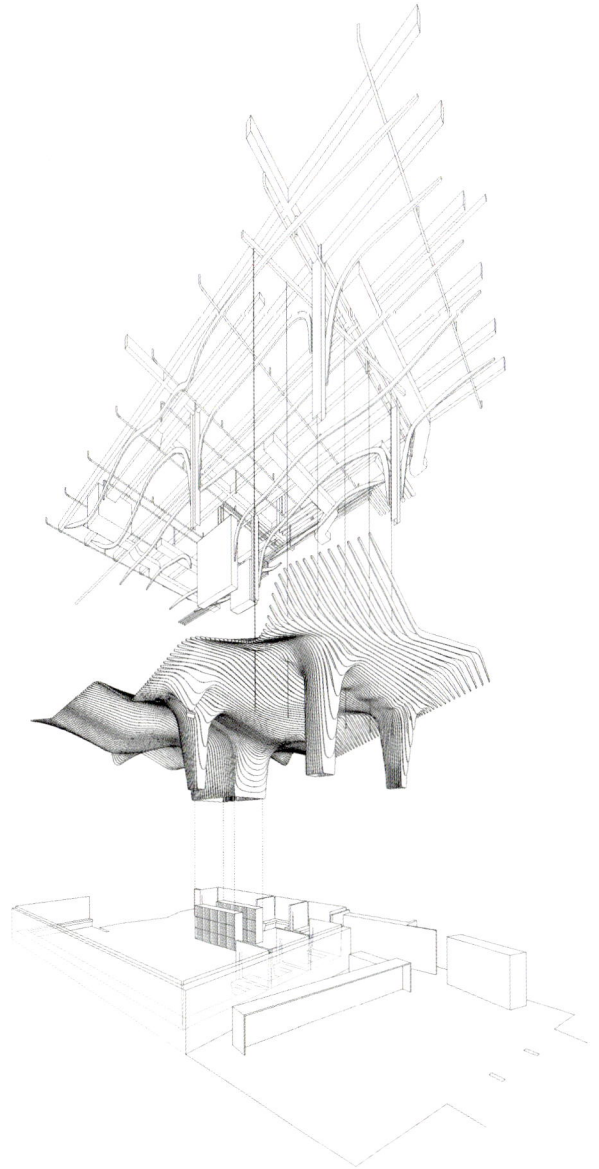

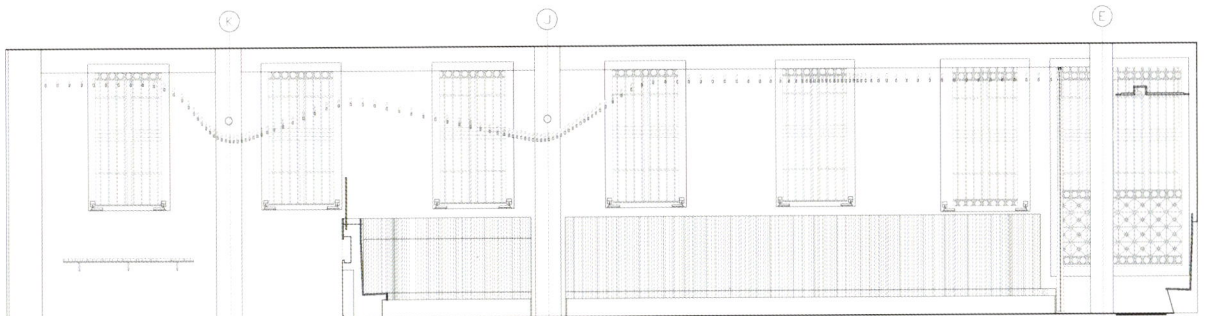

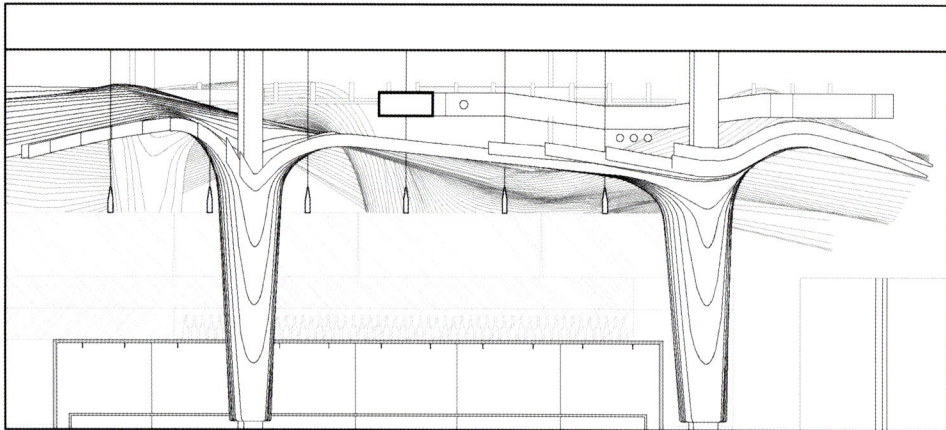

153

Designer: wunderteam.pl
Paulina Stepien
Magdalena Piwowar
Client: Muzeum Sztuki
Photography: Ula Tarasiewicz
Total area: 315 m²

Museum of Art Cafe

The Museum of Art has an avant-garde tradition. It contains 19th and 20th century art, at the same time being contained in a 19th century middle-class palace. In order to modernize the ground floor of the building and adapt it for new functions, it was necessary to pay respect both to the avant-garde tradition, as well as the building's historic function.

We used simple materials, such as plywood, metal and glass, which stemmed from our original concept, i.e. finding a way to emphasize the beauty of the palace's historic interiors, without imitating their style. We employed contrast, juxtaposing the 19th century interior with contemporary design. We were also looking for direct references to the museum as a building and came up with an interior that would resemble a forgotten art warehouse, containing mobile furniture, reminiscent of transport crates used to carry works of art, carts, platforms, etc., as well as the 'crystalline' form of the bar.

The latter was the most difficult element of the entire design, both for us and the carpenters. We wanted to build a structure that resembles a fragment of some mysterious sculpture – a form that grows right out of the floor and walls. At the same time, the bar was supposed to be comfortable and visually attractive. We managed to build a 'broken' crystalline structure out of plywood and lit-up Plexiglas, simultaneously respecting the principles of ergonomics. Oh, and one more thing: the result is a perfect match for the bar stools designed by Konstantin Grcic.

艺术博物馆收藏了19世纪和20世纪的艺术品，其本身就位于一个19世纪的宫殿中，因此在更新一层空间以便适合新功用的过程中，设计师既尊重了历史传统，又体现了前卫的风格。她们使用了如胶合板、金属和玻璃等的简单材料，营造了现代设计与19世纪室内空间的一种对比感。吧台的"水晶"造型是整体设计中的难点，这种构造是对一些神秘雕塑的碎片的一种暗喻。设计师最后采用胶合板和树脂玻璃完成了既舒适又颇具视觉冲击力的构造，并表现了对人体工程学的尊重。

Designer: Andre Kikoski Architect
Photography: Peter Aaron \ ESTO
Total area: 148 m²

The Wright

The Wright at the Guggenheim is designed by Andre Kikoski Architect, an imaginative, award-winning Manhattan-based architecture and design firm. The design solution references the building's architecture without repeating it, and in the process transforming familiar geometries, spatial effects and material qualities. The playfulness of forms and the dynamics of movement through this 1,600 square foot space imbue the design with novelty, subtlety and intrigue, in part through the material palette of the space.

The project is representative of Andre Kikoski Architect's style—inventive, dramatic and highly tactile. Sculptural forms for the flared ceiling, undulating banquette, and torqued bar and communal table are crafted in contemporary materials. They are based on Wright's underlying geometries. The design brings to life a play between these sculptural elements and the architecturally-layered, illuminated materials that invite participation and a sense of delight for all patrons.

'We chose materials and colors for these dynamic forms that are restrained and elegant' explains Andre Kikoski. The design features include: a curvilinear wall of walnut layered with illuminated fiber-optics; a bar clad in a shimmering skin of innovative custom metalwork and topped in seamless white Corian; a sweeping banquette with vivid blue leather seating backed by illuminated planes of woven grey texture; and a layered ceiling canopy of taut white membrane.

Andre Kikoski Architect's design philosophy for this restaurant engages the heightened sense of procession that is essential to the experience of this building—and the dynamic perception of art that it fosters. Surfaces and textures are animated by movement, creating an ever-changing fluid aesthetic that is an essential part of the design.

这家餐厅位于纽约古根海姆博物馆内，因此参照了建筑的设计，并对人们熟悉的几何形式、空间效果和材料特性进行了转化，营造了一个新颖、精巧和令人印象深刻的空间。本项目体现了设计师的风格，即创造力、戏剧性和存在感。天花板、波浪形的长椅和弯曲的吧台及餐桌，这些颇具雕塑感的造型元素都是用低调、优雅的现代材料和颜色来进行装饰的，例如，胡桃木材质的曲面墙上设置了光导纤维；吧台表面以金属制品覆面，其台面是白色的可丽耐；蓝色的皮面长椅后面设置了灰色织物包裹的面板；天花板遮篷采用了白色的膜材。所有这些元素的集合都使食客们有一种愉悦的心情来体验空间。

Designer: Charles Doell
Client: Morgans Hotel Group
Location: Hard Rock Hotel and Casino, Las Vegas

Vanity

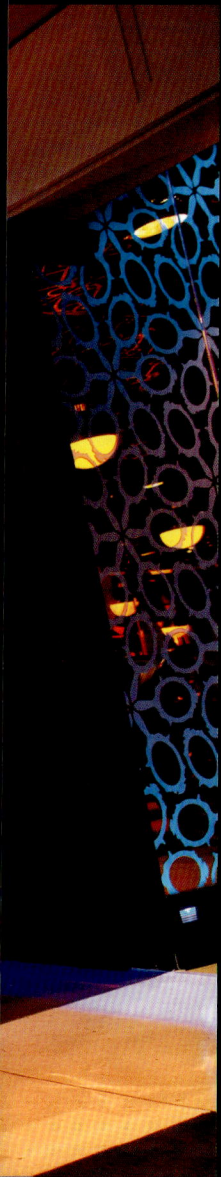

Vanity is an eclectically layered maximalist jewel box of a club. An organized chaos of glinty faceted forms rubbing up against soft cut velvets, deep satins and rich tapestries. It is a mix of saturated jewel tones fused with bronze and gold metallics, antique mirror, rubbed brass and black chrome. References to pearls and hand cut crystal abound. Gleaming textures and honeycombed surfaces wander throughout providing an organic backdrop for the clubs layers of reflective surfaces and parallax views.

The original contact for the project was the Morgans Hotel Group who asked the designers to pitch a concept for the new nightclub at the expanded Hard Rock Hotel and Casino, Las Vegas. They presented a concept that was visually organized by the color, shape and brilliance of jewelry forms. They were especially keen to create an environment that was transportive and would resonate with women as well as men. They paid a great deal of attention to the women's lounge/ bathroom area.

The space is intimate, the dance floor sunken and the entirety of it revolves around a cyclone of crystal and light that rises out of the dance floor and spreads out in all directions above the dance floor. This vortex of glass and light covers 1200 sqft and hovers over the dance floor like a cloud. Comprised of 20,000 pieces of crystal and an equal number of programmable lights the piece is literally the heartbeat of the club.

Simply put, Vanity is a sensualists playground of controlled chaos, precisely like the jewel box fantasies that fuel it. To this jewelry based foundation the designers freely mixed in forms of surface and furnishings associated with a variety of time periods but largely evoking 70's and 80's. Although there is some definitely some fin de siècle about it as well as a huge and very contemporary bi-morphic computer generated lighting sculpture that could only have been produced in the past decade.

This one element is composed of 20,000 programmable LEDs and 20,000 crystals. It rises out of a glass floor and bursts up 20 feet where it spreads and undulates over the entire expanse of the dance floor (1100 sq. ft.). It is capable of streaming video, animation, color changing, messaging etc. Its only limitation is the ability of the programmer to come up with content for it.

The other area of special interest is the women's lounge which features full-height walls of laminated glass with artwork by fashion photographer Miles Alridge. One side has his gold eye and the opposite his lips dripping gold. Between the two is a cloud of 8 multiple armed sputniks with aqua mirrored glass balls and lit white ones. The vanities feature sinks from Jaime Hayon, large individual lit make up mirrors and soft raspberry colored chenille poufs for each sink area. Two sets of 3-way floor to ceiling mirrors add to the complexity and luxury of the space. The rear area of the lounge contains the bathroom stalls which are 'tufted' in a printed graphic for a surreal effect.

In addition to the Miles Alridge artwork the designers have made extensive use of art from Yelena Yemchuk whose brooding shots of women could have been made in the 1940's. These appear in variety of areas, behind glass, stretching 7 meters tall at the dance floor and anchoring the bathroom vestibule and one of the smaller lounge areas.

Morgans酒店集团委托该项目的设计师为其位于拉斯维加斯的酒店提供一个全新的夜店设计概念。借此，设计师将浓烈的色彩、金属、古董镜和熠熠生辉的珠宝造型完美融合，营造了一种感官强烈并能引起所有人共鸣的氛围。空间中四处可见的织物和蜂窝结构的表面也为夜店的多层视角提供了一个有机的背景。一团由2万片水晶和相同数量的灯具构成的"云彩"漂浮在舞池的上空，它可以变换颜色，呈现视频、动画和信息等，一切内容都由控制灯光的程序师决定。而时尚摄影师和艺术家的摄影作品呈现在玻璃后面，被分别放置到女士休息室、门厅和其他更小的地方，这是空间中的另一种视觉焦点。

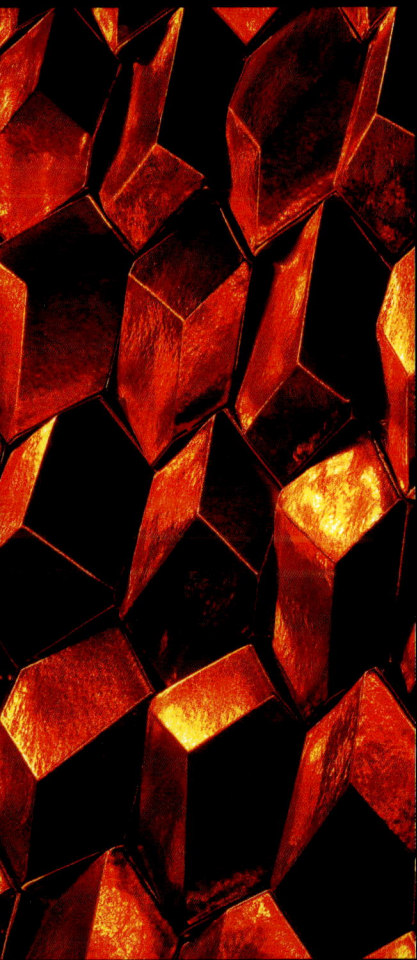

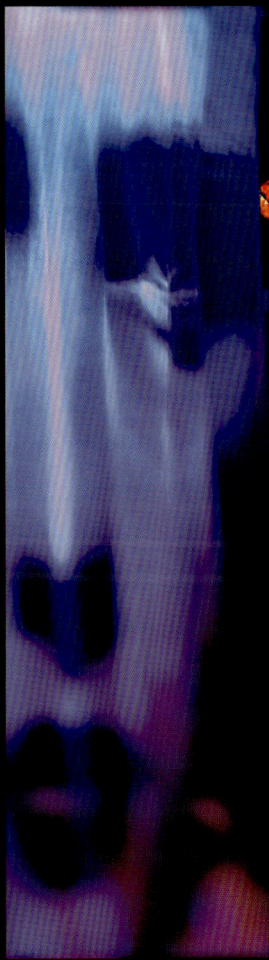

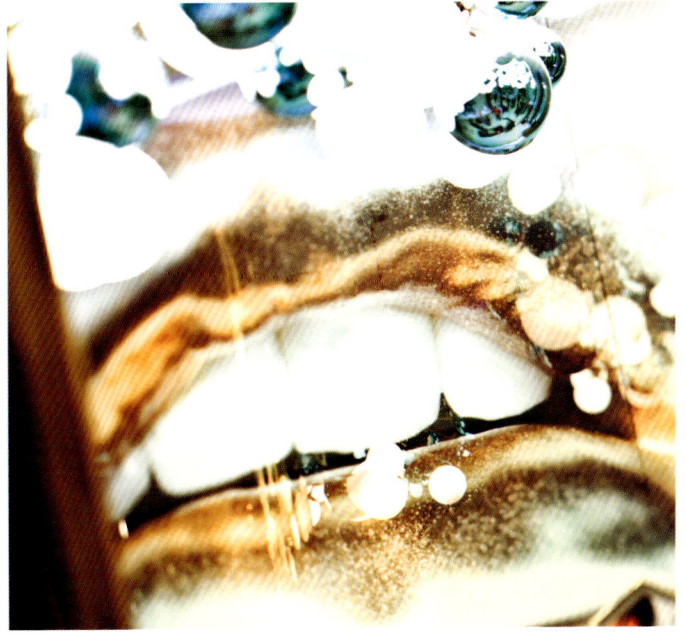

Designer: DarkDesignGroup

Sliver Restaurant

DarkDesignGroup developed a name, logo and interior design project for the restaurant nobiliary cuisine 'Shepka' (Russian 'Sliver'). Supplying Russian folk cuisine was the motive for the birth name of the restaurant 'Sliver', which was the main idea of an interior design project.

Logo of the restaurant contains one of the main elements of interior design—a pot that was used in Russia for cooking. All elements of design and corporate identity of the restaurant support the theme of khokhloma painting—Russian folk decorative painting. In the middle of the dining room there are 4 huge pot-lamps, of which a pair of raised wood chips create a sort of cooking in an old Russian utensils. Walls are painted by hands under the Khokhloma. Floor is made of tile, which imitates the texture of wood. Interior design project restaurant 'Slive' won first place at the VI International Festival design COW 2009, Ukraine, Dnepropetrovsk.

该项目是一家俄罗斯餐厅。在这家餐厅的标识上可以体现项目室内设计的一种重要元素,即与烹饪有关的容器。室内墙面上装饰的是传统的俄罗斯手绘图案。地面铺装了瓷砖,与木材的纹理保持一致。在餐厅的中心区域设置了四个大型的似烹饪容器一样的灯具,从其上面升起一直延伸到天花板的不规则排列的许多木条是对传统烹饪方式的一种隐喻。

Designer: Outofstock
Location: Singapore
Total area: 58 ㎡

Hatched

Located in a fifty-year old student dormitory building next to the National University of Singapore's Law Campus, Hatched serves up a curious array of egg-inspired dishes and desserts catering to a youthful target audience. The intent behind the design was to create a cozy and fun dining atmosphere which identifies with the restaurant's theme of breakfast and eggs. Another challenge was to accommodate up to 38 diners and a kitchen into the 58 square-meter space.

Outofstock divided the dining areas into two parts—a central dining area which features long communal dining tables reminiscent of summer camps, and a fun-wall area where customers dine in front of a large blackboard-painted wall which allows the guests and staff to doodle or leave messages. A long bar counter features a facade consisting of oak, maple and beech veneers and accompanying bar stools also feature the same three solid woods.

The play with natural light is a key ingredient which characterizes the space and Outofstock designed oak veneered panels with egg-shaped cut outs to filter light in from street-facing glass windows. These 'egg panels' also serve as an identifiable frontage for the restaurant.

The use of guava tree branch off-cuts sleeved over bulb holders adds a tinge of farmhouse charm and all the exposed light bulbs are connected to dimmers so as to regulate the brightness needed during the day and night. Although there has been much phasing-out efforts of incandescent light bulbs in recent years, the rich, warm glow and character of near obsolete carbon filament bulbs Outofstock found at a specialty bulb supplier was difficult to resist. Outofstock chose to expose the original concrete ceiling of this historical building and highlighted its textural qualities with lighting.

Hatched位于一座具有50年历史的学生宿舍建筑内，主要经营项目是为年轻人提供餐品和甜品。因此，本项目的目标是营造一个舒适有趣的餐厅环境，并与餐厅的主题，即早餐和鸡蛋相呼应。另一个挑战是在58平方米的空间内容纳38位就餐者和一间厨房。设计师将空间一分为二，主要就餐区摆放让人联想到夏令营的长条餐桌，另一区有一整面黑板，可以涂鸦和留信息。自然光从橡木饰面的薄板上鸡蛋形的切口中投射进来，而稍显过时的碳丝灯则为室内空间添加了别样的韵味。

Designer: Outofstock
Location: Singapore

Bella Pasta

Italian chef Antonio Manetto commissioned Outofstock to design the interiors of his restaurant Bella Pasta, located at 30 Robertson Quay, Singapore. Robertson Quay is located along the Singapore River and is well known for its calm atmosphere and restaurant culture. Chef Manetto was originally from Napoli in southern Italy, and he wished for Bella Pasta to have the carefree yet homely emotional quality of pier-side restaurants in his hometown.

The result is a breezy, relaxed space with nuances of Italian architectural elements such as cobblestone floors, semi-private nooks, arched windows and doorways. Outofstock also specially designed and custom-made the chairs, tables, lighting fixtures and shelving solutions.

A feature of the restaurant is the minimal roman arches which frame the wine rack, open kitchen and pasta display shelves. The ceiling lamps are inspired by dining plates, turned solid oak discs with a crown mirrored bulb emitting soft, indirect light.

Bella Pasta的主人是一位来自于那不勒斯的意大利厨师，而餐厅的位置位于新加坡河畔以安静的就餐氛围和餐饮文化而著名的区域内，因此他希望体现一种休闲又温馨的空间品质，与其家乡所带给人的感受一样。设计师以意大利的建筑元素，如圆石路面、半私密的角落、拱形窗和门为业主打造了他的理想空间，并为他专门设计了桌椅、灯具和搁板。框定酒架、厨房和展示架的罗马式的小拱门是餐厅的一大特色，而屋顶灯具的设计灵感来源于餐盘，它们投射出柔和的间接光源。

Designer: STUDIOUNODESIGN
Photography: Simone Rocchi
Location: Vinci, Italy
Total area: 100 m²

Scarlett Cafe

Scarlett Cafe is a 100 square meter bar and restaurant with a cool interior environment that spread serenity and a touch of elegance in Vinci, Italy. The idea of STUDIOUNODESIGN was based on the contrast between form and decoration, intersecting and combining the basic forms of minimalist aesthetics and square with sparkling floral decorations that are reflected and transformed on the floors.

The floor color only black highlights even more shapes and colors in contrast, emphasizing the details and stylistic choices of objects of decoration. The two environments — bar and restaurant — seem to overlap each other with a unique design that creates a sense of continuity and harmony.

Flowery carved figures on the wooden marks, back-illuminated dark glasses, with programmed LED light, create a cozy warm sensation to the center-placed bar.

The square-shaped bar allows the public to approach the internal hall through two huge openings. The two crystal blades give a sense of fragmentation that unites with the origami lamps, recalling the whole sense of the Scarlett Cafe.

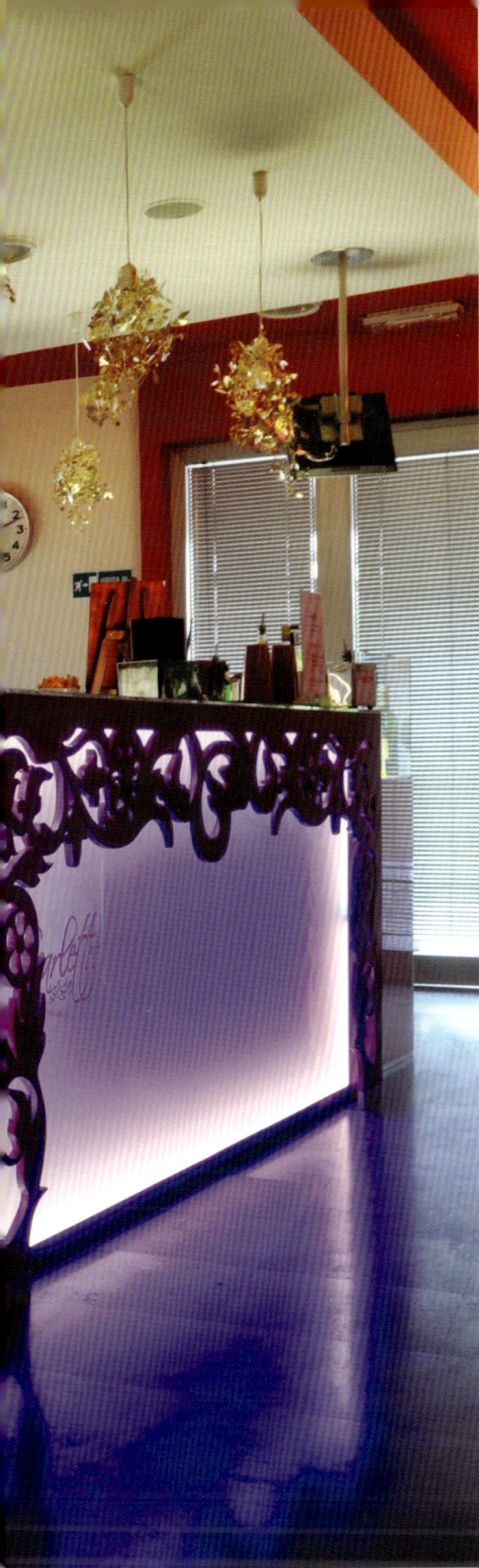

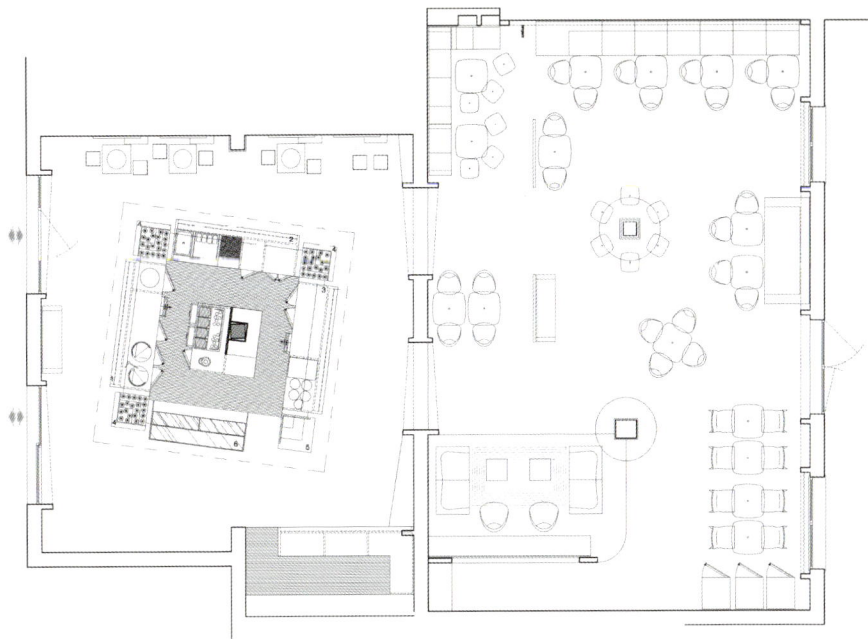

该餐厅与酒吧项目传递了一种宁静和优雅的氛围。设计师的设计灵感基于造型与装饰之间的对比，他们将极简美学元素与熠熠生辉的花形装饰完美融合在一起。空间地面的颜色只有黑色，因此更加突显了相互对比的造型和色彩，强调了装饰细节。酒吧与餐厅的环境似乎彼此交错，营造了一种设计上的延伸感和和谐感。花形装饰图案出现在吧台和玻璃上，在LED灯光的映衬下，为酒吧营造了舒适温暖的感觉，而设计师选用的形似日式折纸工艺的灯具与整体氛围相得益彰。

Designer: Jonas Wagell
Client: Stockholm Furniture Fair
Photography: Jonas Wagell D&A
Total area: 300 m^2

Design Bar

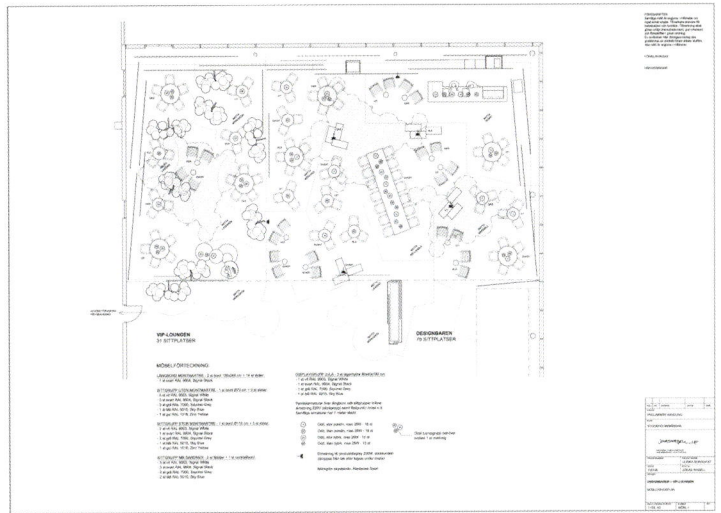

Trade fairs are temporary constructions quickly built to be torn down shortly after. Instead of creating an installation which attempts to be more than this, the Design Bar of year 2010 aims to embrace the temporary by creating a space which is influenced by stage design and graphics rather than polished architecture.

This year's Design Bar is commissioned to Swedish architect and designer Jonas Wagell who will — by small means and simple measures — create an expressive space with strong character. There will also be a VIP lounge which shares the same conceptual idea, but with different look and feel.

The bar area will serve light foods and provide a surrounding where you can sit back and enjoy a drink or coffee, while the secluded VIP area will offer an undisturbed seating for meeting and chats. The conceptual theme for the Design Bar and the VIP Lounge is Forest and Industry — a tribute to raw materials, craftsmanship and refinement, which constitutes the backbone of the furniture industry.

The Design Bar will be furnished with new products by Jonas Wagell — The 'Montmartre' furniture set from Mitab, 'Mr Gardner' outdoor easy chairs from Berga Form and the 'Cage' steel baskets and the 'Odd' family of pendant lamps, bowls and vases from Hello Industry.

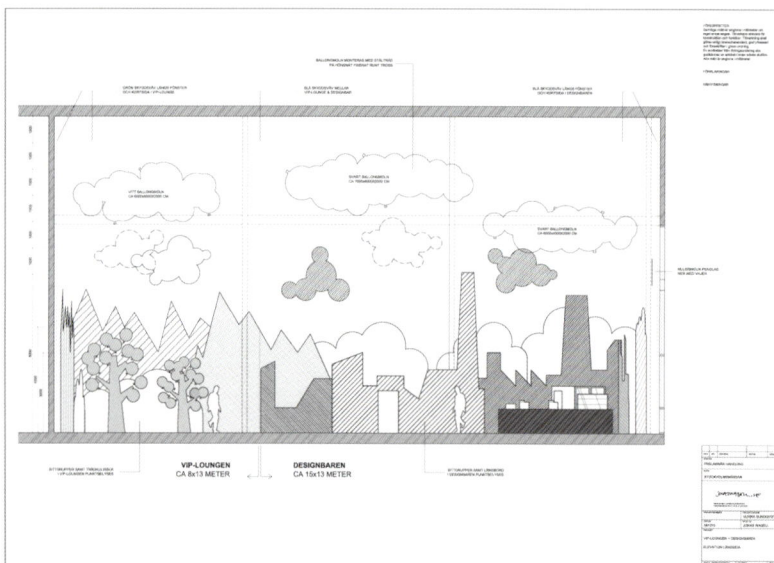

该项目是为斯德哥尔摩家具展览会所进行的设计。设计师从舞台设计和平面设计中获得了灵感，遵循项目临时性的特点，以简单的方法营造了一处令人印象深刻的空间。该项目由酒吧区和VIP区两个部分组成，参观者可以在酒吧区坐下来，享用一些食品、饮料和咖啡，而VIP区为参观者提供了相互交流的场所。两个不同区域的设计主题分别为森林和工业，强调了原材料和手工艺，而这正是家具生产的基础。

DESIGN BAR
the industry

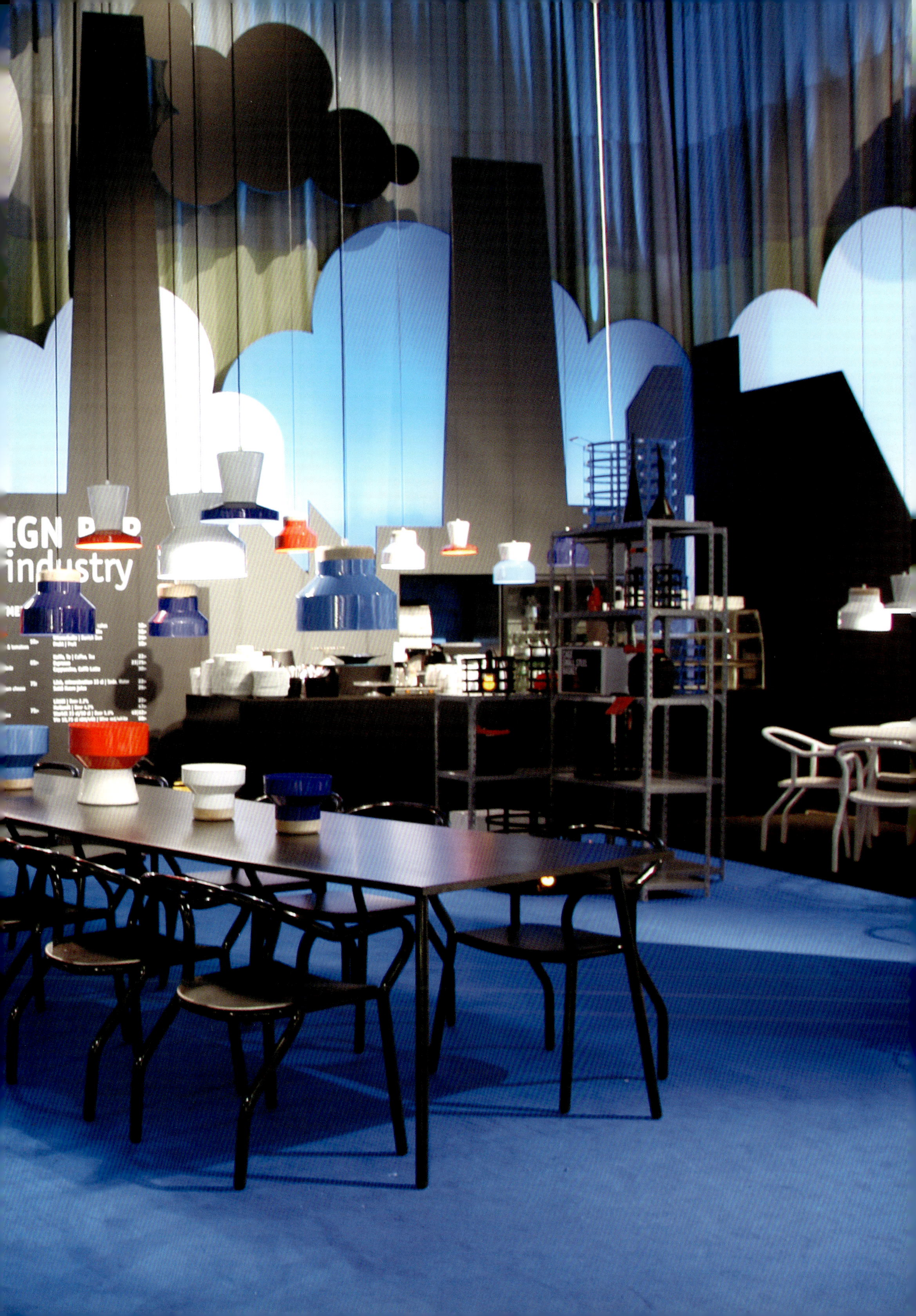
industry

Designer: nendo
Photography: Jimmy Cohrssen
Location: Tokyo

Tokyo Baby Cafe

The design for a 'parent and child cafe' on Tokyo's Omotesando, for parents to enjoy being out with small children without worry about those around them. The cafe is fully stocked with picture books and toys, and includes a playroom, private rooms and separate spaces for nursing and changing diapers. Wide aisles make it easy to move around with a stroller, and light switches and door handles are placed high up to keep children from using them.

The cafe is designed to be enjoyed by two very different sizes of users, 'parents' and 'small children', so the interior plays on this difference in scale. They also see the world through different eyes. Take a table: adults live their lives aware of tabletops, and the things placed on top of them. But children see the table's underside. A table's legs can look like pillars, and the reverse of the tabletop is like a roof. The cafe's 'absolutely huge' and 'absolutely tiny' furnishings take advantage of these two perspectives, the adult's and the child's.

A nursing sofa becomes a playroom when blown up on a massive scale, and a diaper changing table when shrunk to minuscule proportions. Big windows pair with small ones, and big light bulbs with small ones. The floorboards vary in size, and the undersides of tables, where parents eyes don't reach, hide pictures of parent and baby animals. In fact, 'parents and children' can be found all around the cafe, ready for their parent and child visitors.

这是为父母和孩子共同设计的一间咖啡店，使父母既可以享受休闲时光，又无需担心孩子的安全。咖啡店中放置了图画书和玩具，包括一间娱乐室、私人房间，以及为看护服务的独立房间。父母和孩子看世界的视角不同，以桌子为例，父母看到的是桌面和上面放置的东西，但在孩子的眼里，桌底成了天棚，上面可以装饰可爱的照片，而桌子腿成了立柱。设计师正是利用了这些不同，将"绝对的大"与"绝对的小"在这一空间中尽情运用并统一起来。

Architects:Waltritsch a+u
Location: Gorizia,Italy
Project Team: Dimitri Waltritsch and Federico Gori,
Leonardo De Marchi, Cecilia Morassi
Project Area: 500 m²
Photography: Marco Covi. Trieste (© Dimitri Waltritsch)

Casiraghi Gorizia Mediatheque

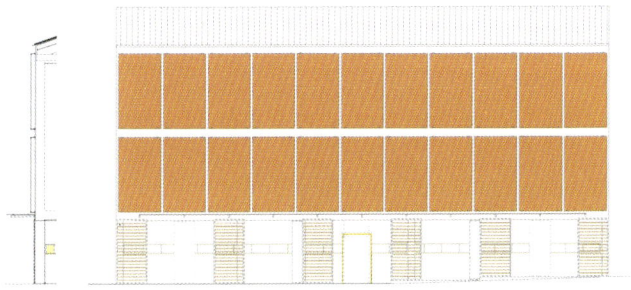

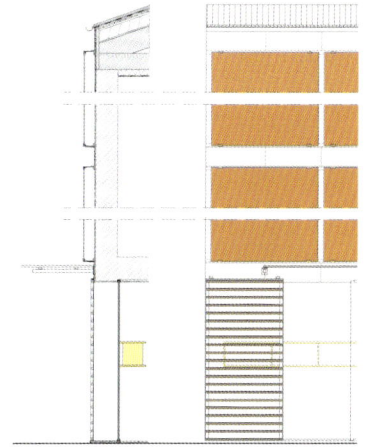

The new Mediatheque is part of a larger complex named Casa del Cinema—Home of the Film, which includes the Kinemax multiplex, several associations dedicated to the cinema culture, the DAMS Cinema section of the Udine University, and finally the Mediatheque. One place, located between the city main square and the castle hill, which gathers commercial, cultural, educational and promotional activities dedicated to the film culture.

This combination of different activities is obviously quite unique, and particularly important for the small city of Gorizia. The Mediatheque stands on the ground floor between the street and one internal passage, so it has two entrances, facing the city as well as the University. The simple plan layout divides the space into three main areas open to the public: the newspaper and magazine hall, the study space and the video room. Behind the reception and reference point, which is visually connected to both entrances, there are separated rooms as storage and one office. All spaces are bound by book and media shelves at full height. One shelve line is marked by a strong color, different for every area, providing specific identity. The same colored shelve line defines the glass facades as well, becoming a communication vitrine, where you directly expose new arrivals, or organize a small exhibition directly facing the public street. The newspaper and magazine area have a custom-designed star shape reading table and a cross shape information counter, and is thought for informal gathering.

mediateca provinciale
di gorizia
"ugo casiraghi"
goriška pokrajinska
mediateka

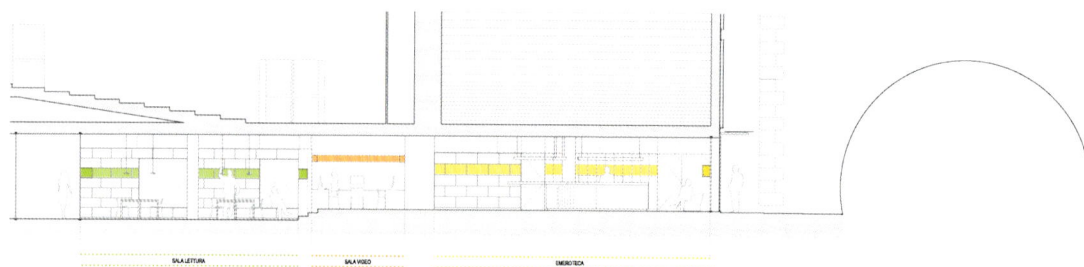

SALA LETTURA   SALA VIDEO   EMEROTECA

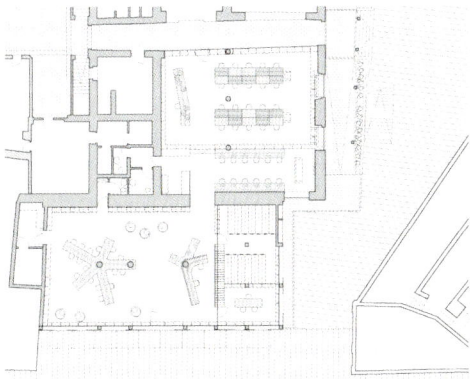

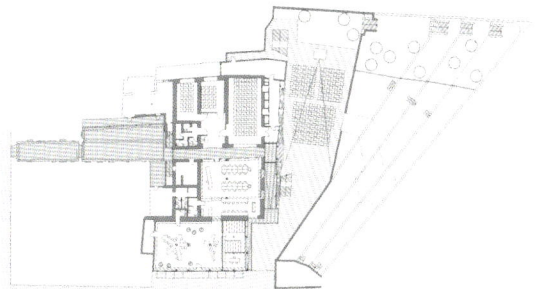

The tables in the study room can be reorganized in order to host reading evenings or presentations. Part of the project is the new facade on the public street as well. A series of colored glass panels on the higher part of the facade are facing the built and natural context of the historical city heart. The dialogue with the surrounding buildings goes through the use of the typical color palette of the building render, and the slight and not intrusive reflection of the surroundings provided by the colored glass. This allows the context to be dilated into the Mediatheque building facade: a 'form of transit' of the everyday life.

这个媒体中心是一个大型综合项目的一个组成部分。这一综合项目位于城市主广场和城堡山之间，融合了与电影相关的商业、文化、教育和宣传活动等。这种多功能的特点是十分特别的，对于戈里齐亚这座小城来说也是非常重要的。媒体中心位于街道和内侧通道之间，其两个入口将城市与大学连接起来。报刊室、学习中心和视频中心构成了媒体中心内部三个主要的空间。书架装饰了强烈的颜色，不同区域以不同颜色区分，突显了空间的识别性。定制的星形阅览桌和十字形信息台为举办活动提供了设施。而外立面上的彩色玻璃印衬出周围的环境，使设计方案与城市文脉相得益彰。

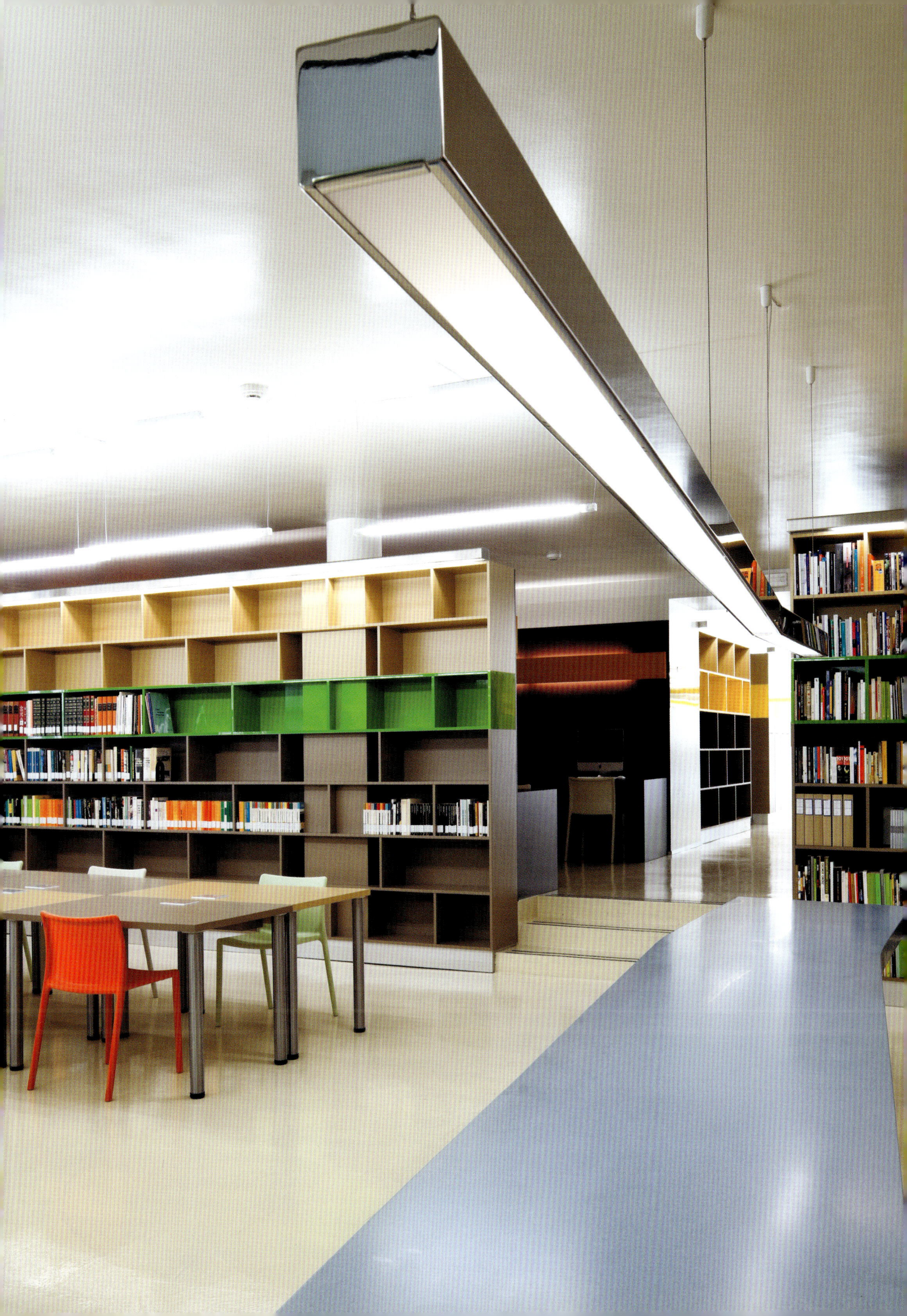

Designer: Stylt
Client: Reval Hotels
Photography: Erik Nissen Johansen
Location: St Petersburg

Reval Hotel Sonya

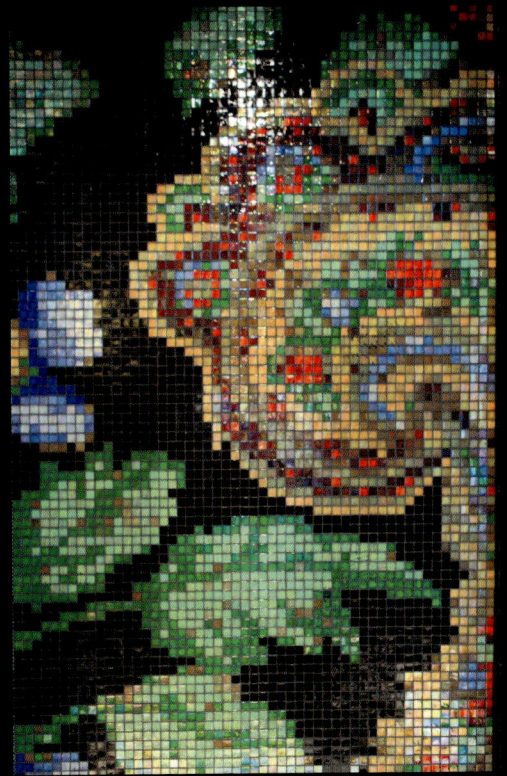

It's 150 years since one of the literary world's most well-known novels was written. But it's only now that you can stay at the hotel. Fyodor Dostoyevsky's classic 'Crime and Punishment' is the basis on which Russia's hottest new hotel has been built, the Reval Hotel Sonya in St Petersburg. When the Baltic hotel chain Reval Hotels wanted to make inroads into the Russian market and create Russia's newest prestige hotel, it employed the services of Stylt, a Swedish firm of architects and identity consultants, which specializes in storytelling. Instead of copying conventional Western luxury Stylt opted to tie the city and its historical heritage and gathered inspiration from a classic 150 year-old novel.

'Travellers' behaviour has changed. People are no longer satisfied with standard hotels that could be anywhere in the world. Nowadays people want real, relevant experiences that provide a taste of the destination while at the hotel. We always start our projects by creating a story and writing a storyline that's then used by our architects and designers. A good story makes hotel management and staff feel involved and also acts as an effective marketing tool,' says Erik Nissen Johansen, creative director at Stylt.

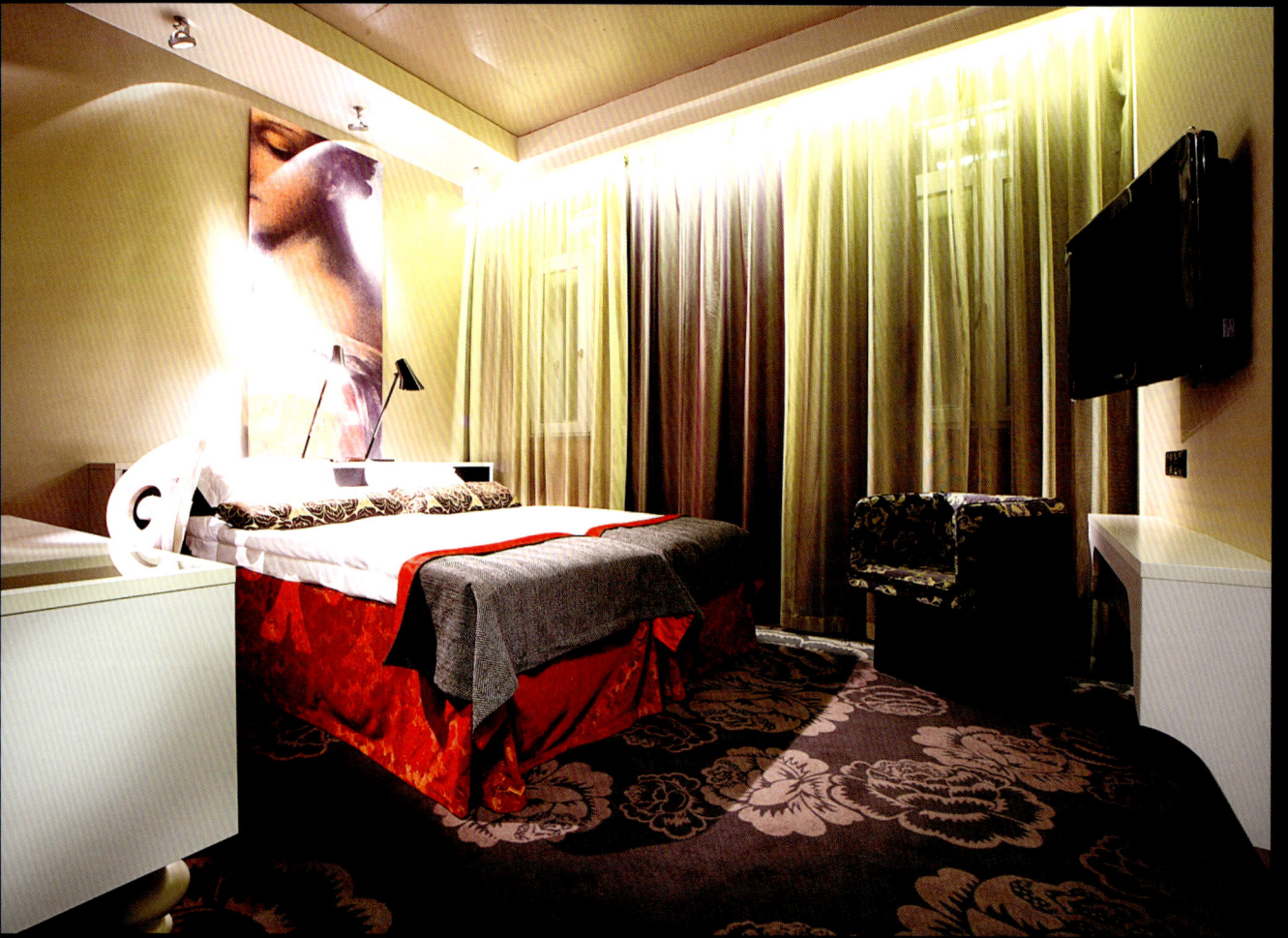

Designer: Blacksheep
Photography: Gareth Gardner
Location: London

Novotel Tower Bridge

Blacksheep created a core concept for this unique location, called 'Liquid History', referring to the Thames, and its associations with journeys, destinations, views and iconic cultural and architectural landmarks. This would translate into both an abstracted wave pattern unifying the newly re-drawn areas of the ground floor scheme and into the idea of views through from one area to another at all times. 'Snapshot London' is a second theme, with famous landmarks represented in specially commissioned black and white photographs. These are used as full-height panels in the scheme and also as doors, including a 'secret door' from the business breakout area into the restaurant space.

The bar is linked to the reception area through the same color palette: seating is in black, ice blue and burnt orange (with some seating in two tonesan ice blue back and orange inner back and seat pad), but contrast is provided through the use of dramatic black floor tiling to add refinement and create a more 'moody' relaxing space. A new rear bar display was created with timber panelling and mirrors an existing bulkhead was illuminated with spotlights.

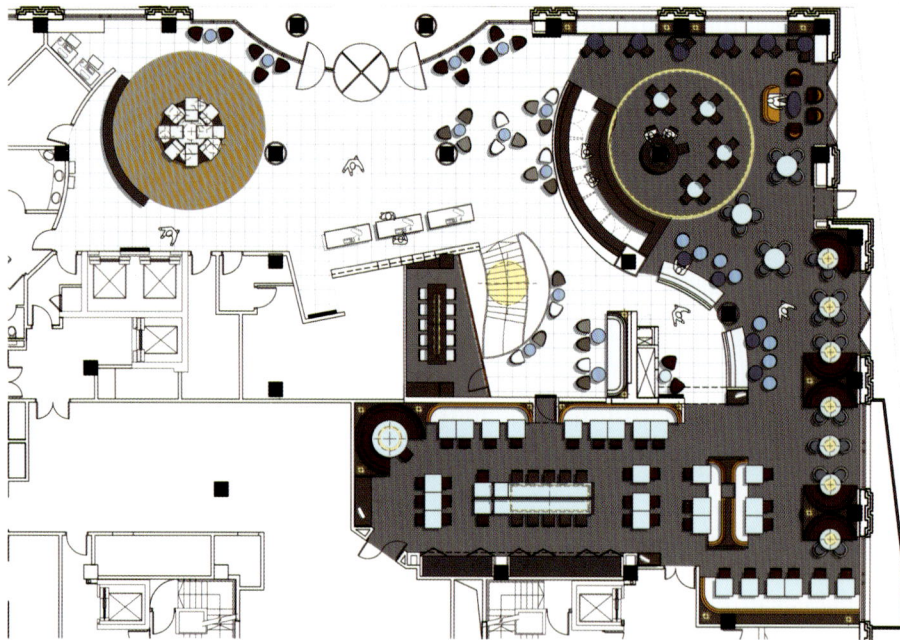

The restaurant area was a long tunnel space with no natural daylight. Large angled mirrors, slightly offset from the wall, are placed on the upper section of opposing walls to help create views, light, reflection and movement. The restaurant features hot and cold buffet areas and a variety of seating for different user groups. Pendant lighting along the window elevation and over the servery draws passer-by attention into the space.

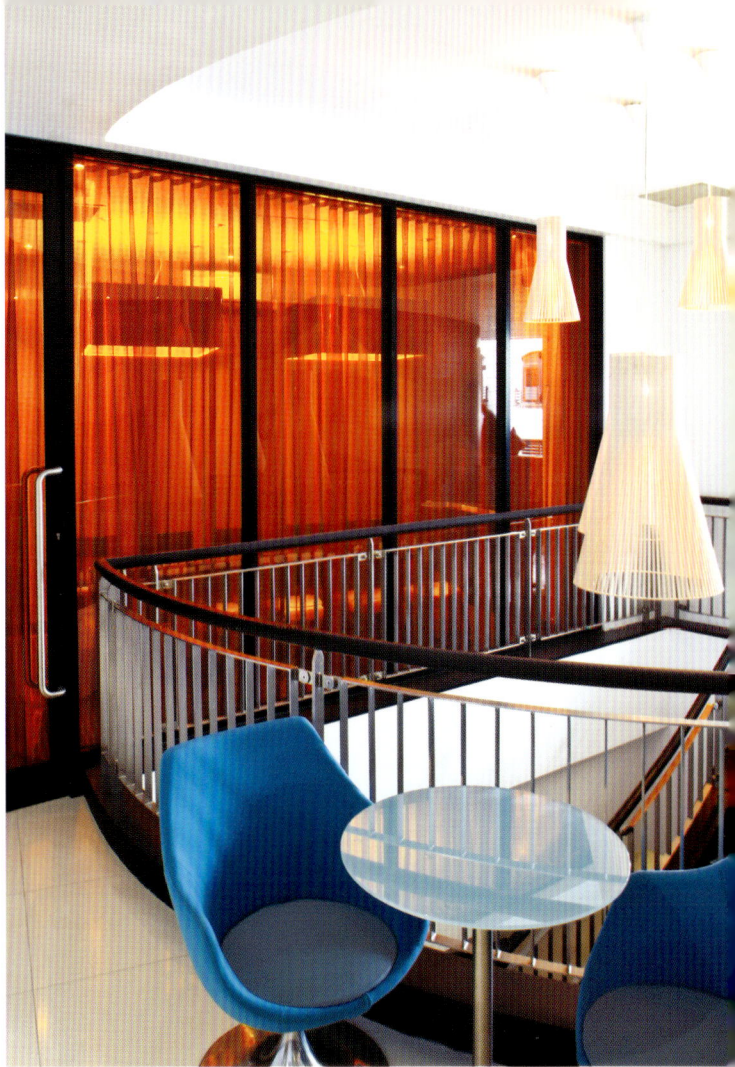

为了回应泰晤士河与旅程、目的地、景致、文化和建筑地标的联系，设计师为该项目确定了两个主题，即"流动的历史"和"伦敦掠影"。前者体现在室内空间抽象的波浪形图案上，而后者体现在门板上的黑白照片上。酒吧与前台的色彩方案是一致的：黑色的、冰蓝色的和橙色的座椅，以及与之相对比的黑色地板，营造了一种颇具氛围的休闲空间。为了解决狭长的餐厅区域无法引入自然光的问题，设计师在墙面上安装了多面略微倾斜的镜子，这有助于反射光线、扩充空间视觉和营造动感。餐厅的座椅大小不一，可以为不同群体的顾客提供餐饮服务，而沿立面摆置的吊灯又将路人的目光吸引到室内空间中。

Interior Designer: Jordi Galí & Estudi
Client: Barcelo Hotels & Resorts
Photography: Jordi Miralles
Total area: 10 000 m²

Hotel Barcelo Raval

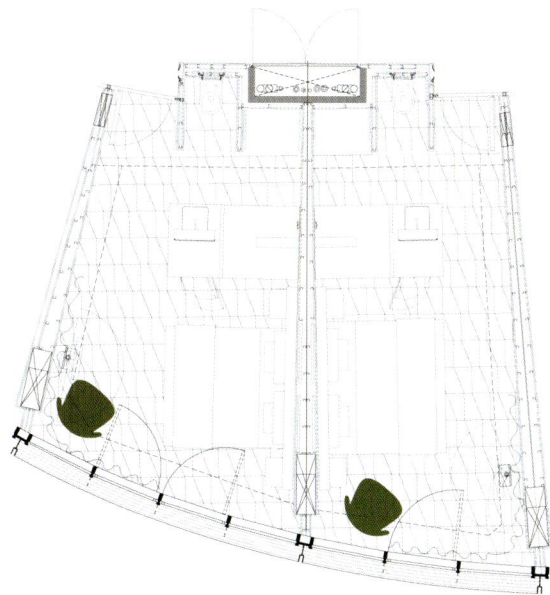

This renovation has raised everybody's attention and it has become the central gem of the city. It is located in the infamous Barrio Chino which has given the area a whole new meaning. This 35 million Euro project was brought to realisation by the architect Jose Maria Guillen White, of the Barcelo development group. Its elliptical shape is what makes it both unique and challenging for a hotel interior. In order to make it functional the architect has raised a central core of concrete, as a backbone, and then some side pillars, so that the rest of the area remains clear. Clear for grand semicircular dining spaces, over-scale lighting and seating forms that have been custom designed to flow with the architecture of the exterior shell making the interior a guided journey to collective items.

Semicircular edges and forms are carried into the interior structure of the bar which is reminiscent of something out of a Sean Connery Bond Movie. As his martini the circulation corridors to the bedrooms are shaken not stirred. The design language is carried throughout the interiors by exquisite cylindrical room signage and projecting floor lighting. One of the most dynamic features in the bedrooms is the elliptical facade iron cladding which is portrayed as a reminiscent decorative element of the 70's.

这一项目是由建筑师实现的，其椭圆形的外形为室内设计营造了与众不同的空间，但也是对设计师的挑战。位于中心位置的混凝土柱在满足功能需求的同时解放了室内的剩余空间，因此设计师得以借用半圆形的餐饮空间、大尺度的灯具和定制的座椅等，与建筑外观相互呼应。半圆形这一设计语汇在客房和灯具的设计上继续沿用，而客房中最具动感的是铁制的外立面覆层，它让人们回忆起20世纪70年代的装饰元素。

Project: Hard Rock Hotel
Designer: Mark Tracy \ Chemical Spaces
Photography: David Marquardt
Location: Las Vegas

Miami Blue Suite

The intent of this design was to create a party suite with a modern, airy, Miami feel. The primary feature is a giant C-shaped bed lined with blue, imported Italian tile. LED rope lighting and mirrored toe-kicks beneath the bed create the illusion that the bed floats above the floor. Laid-out like a party lounge, a solid-surface DJ table in 'Milk Glass' foregrounds the second most noticeable feature: a wall of turntables and jam boxes framed in LED-lit alcoves.

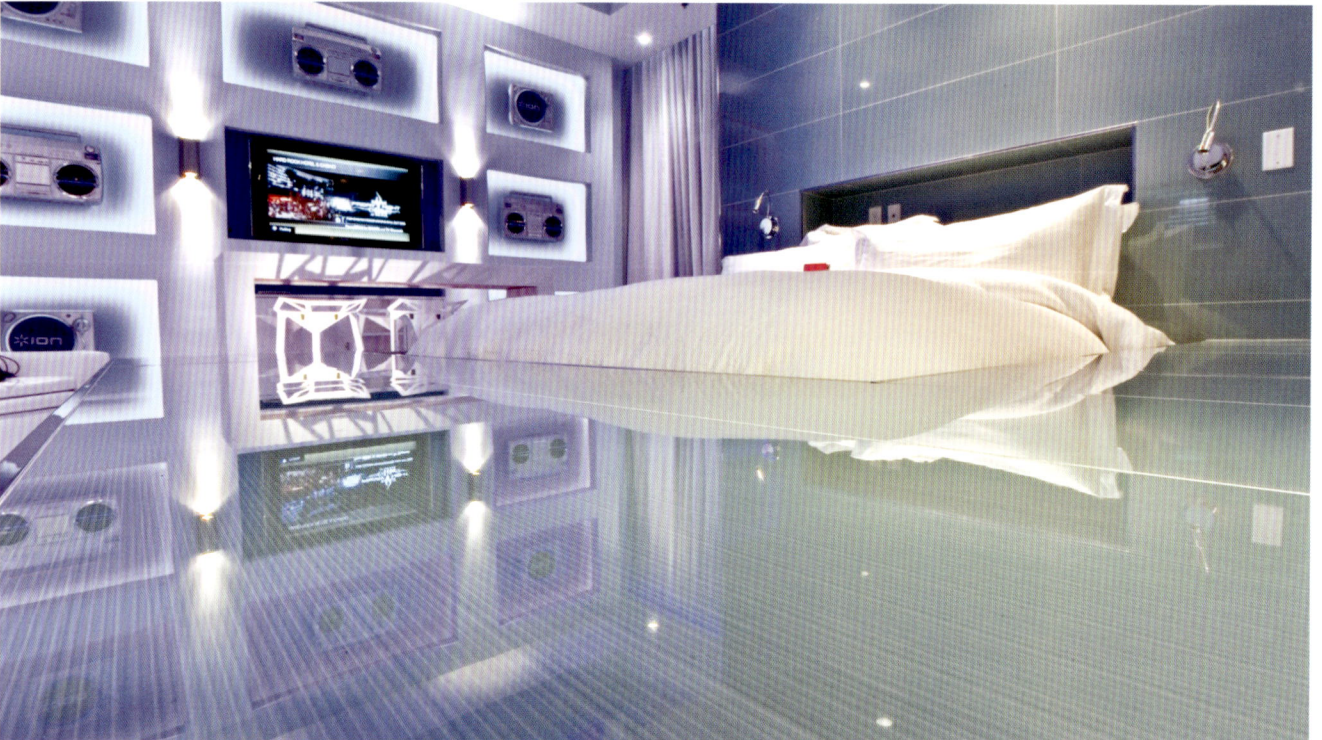

Additional features include custom-designed minimalist furniture, a custom, color-matched rug produced in Sweden, floor-to-ceiling sheers, sconce lighting, and mirrored ceiling niches. The primary challenge was a demanding timeline, having been hired late in the construction process.

设计目标是要创造一种现代的，具有迈阿密感觉的派对式套房。此套房的焦点是一张巨大的C字形的床，其周围装饰着进口的蓝色意大利瓷砖。床底部周围放置了LED灯带，因此看起来床好像是悬浮在地面上一样。套房的另一个焦点是DJ桌后面的墙壁，唱盘和录音机放置在由LED照亮的壁龛内。除此之外，定制的极简家具、瑞典制造的地毯、宽大的薄纱窗帘、玻璃壁灯和镜面天花都是空间中的设计亮点。

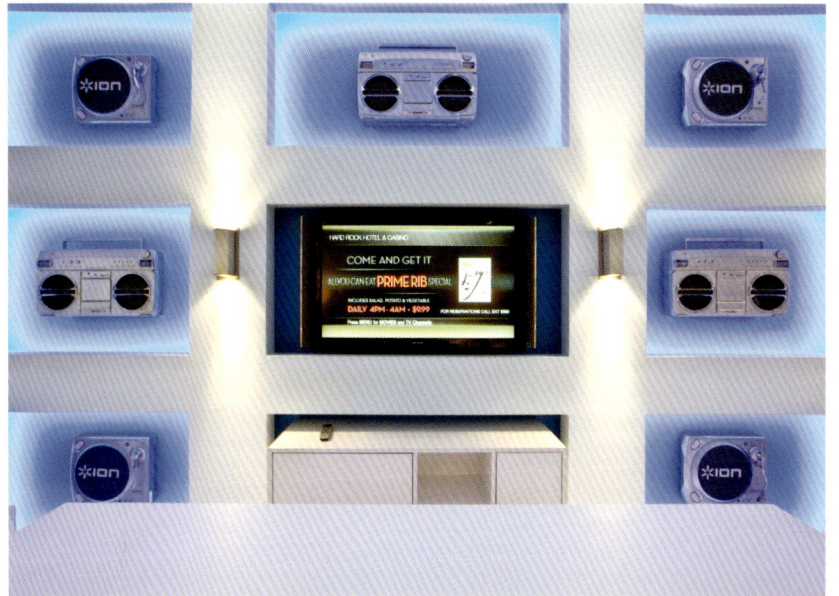

Project: Hard Rock Hotel
Designer: Mark Tracy \ Chemical Spaces
Photography: David Marquardt
Location: Las Vegas

Tree Line Suite

The concept of this design began with the concept of creating a hotel room with the illusion of a two-story tree line in the center. Starting from this feature, the hand-picked colors of tree branches were the basis for the color scheme for the rest of the room. Hardwood floors complement the nature motif. Cream walls of textured tiled anchor each side of the room. To create contrast and clarify that this space is still a party suite, a rich blue band spanning from wall, to ceiling, to wall, divides the room down the center. The blue mural features hand-painted smoke in dark blue, leaves and skulls in metallic gold, and LED rope lighting to illuminate the art.

设计灵感源于在套房中心营造双层树线的概念。围绕着这一点，套房中的颜色都以树枝的颜色为主。硬木地板体现出了自然的特色。为了形成对比，突出派对式套房的风格特点，设计师在房间的中心位置设计了一条分割空间的蓝色缎带，从墙面，到天花板，再到墙面。这条缎带以手绘的深蓝色烟雾图案和金属色的树叶图案为装饰，这些图案在LED灯光的照射下更具艺术气息。

Project: Hard Rock Hotel
Designer: Mark Tracy \ Chemical Spaces
Photography: David Marquardt
Location: Las Vegas

Punk Rock Suite

The intent of the graffiti room was to create a London-style, punk rock party suite. Bands of recessed mirrors and hand-painted lines in every direction, hand-painted wall murals, and spot lighting on the walls and ceiling pack the room with movement and action.

Stone-like grey wall tiles, black hardwood floors, custom modern furniture and a custom, handmade Union Jack rug complete the London underground feel.

设计目标是要创造一种伦敦风格的朋克摇滚派对套房。镶嵌在墙内的带状的镜子、向各个方向延展的手绘红色线条、手绘壁画、灯具和天花板给整个房间带来了动感。如石块一般的灰色墙面砖、黑色的硬木地板、定制的现代家具和定制的手工制作的Union Jack 小地毯完全带给人一种伦敦地下酒吧的感觉。

Designer: Soren Luckins, Dave Williamson, Jules Zaccak
Client: BlighVollerNeild
Photography: Peter Bennetts

MYER

MYER, Australia's largest department store recently relocated their head office to Docklands. BVN Architects commissioned Buro North to design graphic embellishments and signage throughout the space. After a thorough research phase, the designers crafted design which evolves throughout the nine floors; each floor is themed by a specific decade in twentieth century fashion, starting with the 1910's and working up through the building to the 1990's. Wall Graphics were relief routered and signage developed using a feature material relevant to the decade, toilet pictograms were given the same decade relevant treatment creating a subtle and sophisticated interpretation of MYER's heritage.

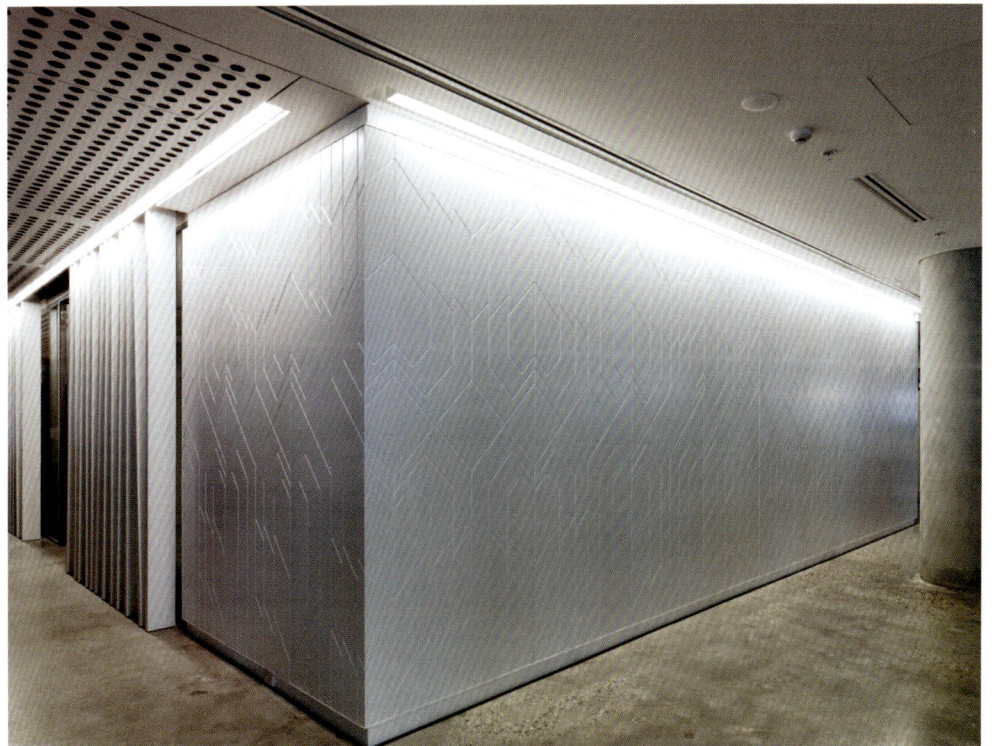

Support Test Lab

该项目是澳大利亚最大的百货商店MYER的公司
总部设计。在完成了对9层空间的研究之后，设
计师决定以20世纪时尚史的一些特殊时间段作
为每一层的设计主题，从20世纪10年代开始，
依照楼层上升，止于20世纪90年代。墙壁上的
图案采用了浮雕法，而指示牌是以相关年代的
代表性材料制作而成的。设计师以一种巧妙的、
雅致的方式诠释了MYER的传统。

Architect: waltritsch a+u \ Arch. Dimitri Waltritsch
Photography: Marco Covi. Trieste (copyright Dimitri Waltritsch)
Total area: 600 m²

Cogeco Headquarters

The City of Trieste is among the world most important port within the coffee trade market. Cogeco is a firm dealing as intermediate between the raw good and the coffee roasting plants. The project consists in the interior renovation of the company siege, underlining the two distinguishing factors which characterize the firm: the worldwide commercial relationships and the fact that Cogeco provides specific knowledge and lab test. These points have been particularly enhanced in the entrance lobby and in the proof and taste laboratory room.

The lobby of the Cogeco siege is characterized by a multi-layered folded wall, an abstracted map where only the parallels of the globe have remained, which hosts a series of 'exotic' coffee names coming from all over the world, and by a big chemistry formula of the caffeine, which marks the corner and gives evidence of the specific knowledge. In such a way, one is immediately transported in a quick ride through the globe, and at the same time given a clear statement of the company's know-how.

The proof and taste lab is the place where the company makes a series of tests on the raw good in order to provide a certificate of quality (roasting, checking dimension and smell, tasting etc), and where most of the commercial deals are made. Coffee sample bags are exposed on a colored shelf marking the perimeter of the room, while one of the proof tables hides the 'spitting pots'. The space is again dominated by a big map folding on this table, a very detailed Goode Homolosine projection of the world, where precise geographic indication about the origin of the most important coffee types are given, adding a visual layer to the proofing experience.

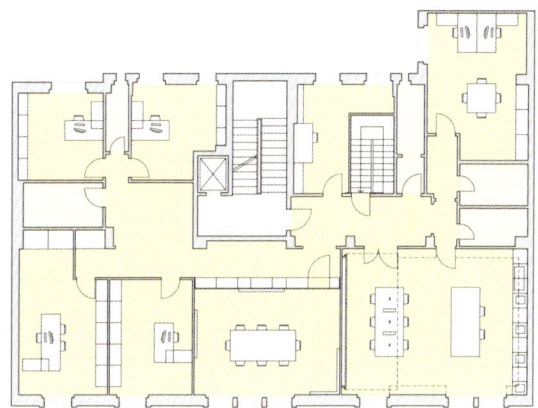

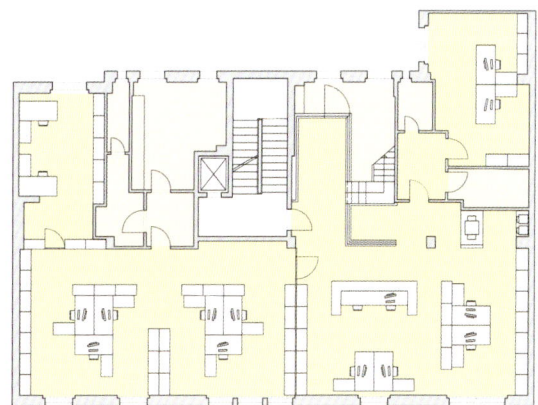

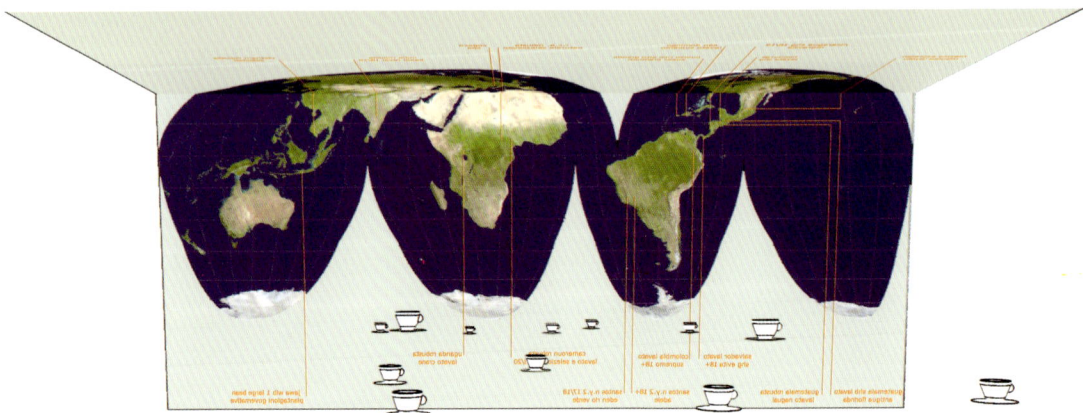

Cogeco公司所在的德里雅斯特是世界咖啡贸易的重要港口城市之一，因此项目设计的焦点是突显全球范围内的商业性和公司的专业服务。Cogeco的前厅以多层折叠墙面和一张抽象的世界地图为特色：地图上只保留了几条平行线，标注了来自于世界各地的咖啡的名称；而墙角装饰着放大的咖啡因化学分子式，再一次强调了公司的专业性。在这些设计元素的引导下，来访者一方面可以纵览全球市场，同时又了解到公司的主要业务。

Designer: emmanuelle moureaux architecture + design
Client: Nakagawa Chemical CS Design Center
Photography: Hidehiko Nagaishi
Location: Tokyo

Kaleidoscope

The Nakagawa Chemical CS Design Center displays as many as 1100 colors. Taking inspiration from a kaleidoscope, the designer used these colors to create stunning office spaces. This exhibition focuses on one color—yellow, red, green, blue or black—at a time. Every month, the space is designed with a different color, changing hues like a kaleidoscope: a rediscovery of colors that ordinarily pass unnoticed in everyday life. If you look closely at the pictures you can see it isn't just one or two glass windows. There are multiple layers of glass panels that are carefully layered to give that kaleidoscope effect. The exhibition space changed every month over a period of five months.

kaleidoscope

CS Design Center展示了约1100种色彩，因此设计师以万花筒为灵感，使用色彩为办公空间营造了一种令人惊奇的效果。这一设计每次以一种色彩，即黄色、红色、绿色、蓝色或者黑色为主题。每个月，空间的主色调就会变换一次，这就像万花筒一样，目的是重新审视在日常生活中被忽略的色彩表现。如果仔细观看图片，人们会发现设计师使用了精心布置的多层玻璃板，借此营造出如万花筒一般的效果。

CONTRIBUTORS

waltritsch a+u \ Arch. Dimitri Waltritsch

waltritsch a+u is operating since year 2001 in Trieste and is directed by Dimitri Waltritsch. The office activity is spanning between architecture, interior design and urbanism. Among realized projects the cultural and educational center KBcenter in Gorizia and the urban plan Casanova for the extension of the city of Bolzano with 1000 new homes and services. Current projects include the Gorizia Provincial Mediatheque, the K Cultural Center in Trieste and the CasaDelser Product and Image Center in Udine. Projects and buildings are often published throughout Europe and Asia. Dimitri Waltritsch is teaching Architectural Design at the Faculty of Architecture of the Trieste University, has been teaching urban design at the Ferrara Faculty of Architecture, has been juror for the Öfsterreicher Bauherren Preis, visiting lecturer at the Columbia University in New York, the Berlage Institute in Amsterdam, the Universities of Venezia and Firenze and the Haus der Architektur in Graz.

Website: www.studiowau.it

Chikara Ohno \ sinato

Architect. Born in 1976 in Osaka. Established sinato in 2004. Handled various projects relating to houses or business facilities, nominated as top 40 world emerging designers by I.D. Magazine (US) in 2009. Received various domestic and international awards.

Website: www.sinato.jp

Mark Tracy \ Chemical Spaces

Mark Tracy has been designing interiors since 1988 when, as he was studying pharmaceuticals in college, he began painting apartments and creating small pieces of art. Twenty years later, he has accumulated over 200 projects. His body of work is comprised of gourmet restaurants, high-design bars, hair salons, low- and high-rise condominiums and most of all, private residences. Nearly all of Mark's work is in Las Vegas, Nevada. During his time in Las Vegas, Mark has received numerous awards from the Las Vegas design community, including 'Best Designer' in 2008 (Las Vegas Home & Design HEIDI Awards), and 'Most Dramatic Space' in 2009 (NEWH Hospy Awards.)

Jordi Gal & Estudi

Jordi Gal & Estudi opened in 1969 in Barcelona, the company is dedicated to interior architecture and design.

Website: www.jgaliestudi.com

emmanuelle moureaux architecture + design

Emmanuelle Moureaux
1971, Born in France; 1995, Graduated in Architecture, University of Bordeaux; 1995, Received Architect's Diploma from the French Government; 1996, Moved to Tokyo; 2001, Started working as a freelance designer; 2003, Received the Japanese First Class Architect's license; 2003, Established Emmanuelle Moureaux Architecture & Design, Tokyo; 2009, Renamed to Emmanuelle moureaux architecture + design; Associate Professor, Tohoku University of Art & Design; Member of the 'Tokyo Society of Architects'.

Website: www.emmanuelle.jp

298

Gitta Gschwendtner

Gitta Gschwendtner's design consultancy includes furniture, interior, exhibition design and public art for arts, cultural and corporate clients. The studio specializes in a 'tailor-made' approach to design. Each project is carefully researched and every solution is an individual response to the project's particular needs. The designs have narrative at their heart and are derived from careful problem solving rather than styling.

Website: www.gittagschwendtner.com

Rockwell Group

Rockwell Group is an award-winning, cross-disciplinary 200-person architecture and design firm specializing in cultural, hospitality, retail, product, and set design. Founded in 1984 in New York by David Rockwell, the firm crafts a unique narrative and an immersive environment for each project. Rockwell interest in theater has informed much of the firm's work, which ranges from the W Hotel, W Union Square Hotel, and Adour Alain Ducasse at The St. Regis New York; Hall of Fragments, the entrance installation to the 2008 Venice Architecture Biennale; interior work and brand conceptualization for the new JetBlue terminal at New York's John F. Kennedy International Airport; the Children's Hospital at Montefiore in the Bronx; the Kodak Theatre in Los Angeles; Nobu restaurants worldwide; and groundbreaking set designs for Broadway productions of 'Hairspray' and 'Dirty Rotten Scoundrels'. Rockwell Group is currently at work on Canyon Ranch Living in Miami, the set design for the 2009 Academy Awards, the Elinor Bunin-Munroe Film Center at Lincoln Center, the Walt Disney Family Museum in San Francisco, and the Imagination Playground initiative. David Rockwell is the 2008 recipient of the Smithsonian's Cooper-Hewitt, National Design Museum's National Design Award for Interior Design.

Website: www.rockwellgroup.com

Blacksheep

Blacksheep was established in 2002 by Courvoisier Future 500 stars Jo Sampson and Tim Mutton, who quickly established a reputation for stylish, aspirational and award-winning designs, which won them commissions from clients as diverse as Hermes, Voyage, Diageo and Gordon Ramsay. Blacksheep designed London's smash hit nightclub, The Cuckoo Club (a 3-times winner at the London Club & Bar Awards), Euro-elite hangout 'Whisky Mist at Zeta' in the Hilton Hotel, Park Lane; Vendôm in Knightsbridge (also a winner at the London Club & Bar Awards) and Conde Nast Innovation Awards finalist Inamo—the much-lauded, world's first interactive ordering restaurant in Soho.

Website: www.blacksheepweb.com

Koji Yamanaka, Yuji Yamanaka, Asako Yamashita \ GENETO

Koji Yamanaka
1979, Born in Kyoto; 1999, Establish Geijutsu NinjaTai (芸術忍者隊); 2000, Kyoto Seika University, Architecture course Jury's special award for graduation project; 2002, Research student in Kyoto Prefecture University; 2004, Change the name to GENETO; 2007-2009, part-time professor in Kyoto Seika University.
Yuji Yamanaka
1980, Born in Kyoto; 1999, Establish Geijutsu NinjaTai (芸術忍者隊); 2003,

Kyoto Prefecture University, living environment course; 2004, Change the name to GENETO; 2006, Tokyo Institute of Technology, Department of Architecture and Building Engineering, Master's degree; NOW part-time professor in Tokyo University of Science.

Asako Yamashita

1981, Born in Hyogo; 2004, Kyoto Prefecture University, living environment design course; 2007, Konstfack Interior Architecture Course, Master's degree Department Scholarship, Konstfack. Awarded for Final exam.

Website: www.geneto.net

STUDIOUNODESIGN

STUDIOUNODESIGN born from the collaboration between Gabriele Bartolomeo and Simone Nuti with the intention of offer innovating and original ideas without loosing an aware face-off with customers and their requirements. This is a project that embraces architecture, interior design styling of decor object and furniture's complements, always trying to achieve a balance between functionally and beauty. STUDIOUNODESIGN's target is the valorization of the spaces and forms to reach the harmony of an ambient. STUDIOUNODESIGN uses advanced technical resources that permit to reach amazing achievements due to the endless dialogue between the two designers, and mostly with the customer. Professionality, details attention in every step of projection and freshness of the ideas and proposals are the features of this project.

Website: www.studiounodesign.com

Jonas Wagell

Jonas Wagell is a Swedish architect and designer with a background in project management and marketing. In 2008 he was named one of 'the world's 50 hottest young architects' by the international magazine Wallpaper and the Mini House prefab concept won 'Innovation Award 2008' by the Swedish Chamber of Commerce for the UK. In 2009 Wagell founded the design brand Hello Industry and showed its first collection, which Wallpaper announced as the 'Best of Stockholm Design Week 2009'.

Website:ww.jonaswagell.se

Bluarch

Antonio Di Oronzo came to New York from Rome (Italy) in 1997 and has been practicing architecture and interior design for fifteen years. He is a Doctor in Architecture from the University of 'Rome La Sapienz' , and has a Master's in Urban Planning from City College of New York. He also holds a post-graduate degree in Construction Management from the Italian Army Academy. He has worked at internationally recognized firms such as Eisenman Architects, Robert Siegel Architects, Gruzen Samton.

In 2004, Antonio founded the award-winning firm bluarch architecture + interiors + urban planning, a practice dedicated to design innovation and technical excellence providing complete services in master planning, architecture and interior design. At bluarch, architecture is design of the space that shelters passion and creativity. The growing intersection of the arts, science and technology is seen as the opportunity to research and represent human organic interaction. Digital tools and technologies are an integral part of the planning process and a preferred means to implement new approaches to design. Based in New York City, the firm is being recognized for both built and speculative work in both publications and exhibitions.

Antonio Di Oronzo's work has been exhibited at the MoMa, Life of the City (2002); The Van Alen Institute (New York City, 2005); Centro Arquitectum (Caracas—Venezuela,2005). His work has been published in The New York Times; US Weekly; Time Out; New York Magazine; Daily News; Life & Style; Frame; New York Observer; New York Post; Sugar; Boston Herald; City Magazine; New York Newsday; People Magazine; New York Post; Public Culture (cover); Perspective(Honk Kong China); Metropolis, I.D., BOB (South Korea); DeZona (Bulgaria); Shotenkenchiku (Japan, cover); IQD(Italy); Interior Design, Hospitality Design(cover); Boutique Design; Total Lighting (UK); Eigen Huis & Interieur(Netherlands); Edno(Bulgaria); Andmag(Turkey); Quintessentially(Turkey); etc. His work has been included in the following books: Demonstrating Digital Architecture (Yutung Liu; Publisher: Birkhauser, Switzerland); Interactive Design 1.0 (Andrea Rossi, Publisher: Yoll Net, Italy); Best of Club Design (Verbus Editrice, Italy), Echo (Hai Chi Publishing Co., China); Eco-lifestyle (Loft Publications, Spain), etc.

Antonio is a Professor of Architecture at City College of New York, School of Architecture, Urban Design and Landscape Architecture. He also teaches interior design at Parsons School of Design, Department of Architecture, Interior Design, and Lighting. He has taught graduate courses in digital culture and aesthetic, and media design at New School University, Department of Media Studies and Film, and branding and brand management at Parsons School of Design, Department of Design and Management.

Website: www.bluarch.com

wunderteam.pl

Paulina Stepien (2001, Graphic Arts Department, Fine Arts Academy in Lodz) and Magdalena Piwowar (2007, Department of Interior Architecture, School of Art and Design in Lodz) joined forces in 2008 creating a design duo called Wunderteam.pl.

Website: www.wunderteam.pl

Julie Schmidt-Nielsen, Marc Jay, Jenny Selldezn, Linda Vendsalu, Lawrence Mahadoo \ WE Architecture

WE Architecture is a young innovating architecture office, based in Copenhagen, Denmark. Their capability spans from architecture, urban strategies, tangibledesign and utopian ideas. WE Architecture was founded in 2009 by Marc Jay, Julie Schmidt-Nielsen. The name WE architecture is based on the philosophy that architecture is not the result of one person's stroke of genius. They believe that the best results occur through teamwork and trans-disciplinary networks. That is why WE Architecture work across continents as well as across professional borders to enter complex conditions with the best insight and precision. WE create proposals that merge through creative translation of all the information they get from contexts, conditions and programs. WE Architecture strive to push innovative architecture forward to improve the condition of the world. No less.

Website: http://we-a.dk

Andre Kikoski Architect

Andre Kikoski Architect is a Manhattan-based multi-disciplinary design firm that is committed to artistic innovation regardless of budget, genre or client challenge. Our passion for material research, our detail-orientation, and our client-centric approach have won the firm clients in a wide range of categories—from hospitality to arts and culture, from real estate to high-end residential. The firm has been named as one of 'Ten Young Firms to Keep an Eye On' by Oculus, the AIA New York Chapter magazine, and one of 'The New Garde of Ten Designers To Watch' by New York Magazine.

Andre Kikoski Architect's achievements include a nomination from the James Beard Foundation Awards for Outstanding Restaurant Design, a Lumen Award for Lighting Excellence, and the Edwin Guth Memorial Award from the International Association of Lighting Designers.

The firm has completed dozens of luxury town houses, lofts, duplexes and penthouses; residential investment buildings and interiors totaling over 1.6 million square feet; multiple high-end resorts, award-winning restaurants and hotels; and numerous public and cultural projects including prominent cultural venues.

Website: www.akarch.com

Soren Luckins, Dave Williamson, Jules Zaccak \ Buro North

Established in 2004, Buro North is a multi-disciplinary design practice delivering evidence-based solutions that are creative, measurable and meaningful. Led by Design Director Soren Luckins and Wayfinding Director Finn Butler, Buro North's diverse team works across the disciplines of graphic design, industrial design and wayfinding. With an unwavering commitment to quality, Buro North's design and strategy work in tandem in the creation of products, brands, identities, publications, signage, environmental graphics and wayfinding. Buro North's approach is to discover the absolute potential of a project and to resolve their clients' design & communication issues with mature rigour and youthful creativity. It is the diversity of background and specialization within Buro North that facilitates their unique and engaging outcomes.

Website: www.buronorth.com

Stylt

Stylt work based on the idea that the story of the company/brand is a valuable identity asset and the most valuable brand resource. With storytelling as its base, a powerful resource in the form of a tactile brand is created. Based on this, cross-border solutions can be created by architects, interior designers, artists and communicators.

A good story becomes the foundation for clear, attractive and highly distinctive concepts. Stylt has its merits primarily in the experience, destination and hospitality industry and has succeeded in creating a number of very commercially viable concepts. The solutions commercial viability has made the concepts persistent, profitable and brand building. Stylt's core product is providing a new communication tool; Storytelling. Their core competence is to stage a story with a customer-centered and experience-based communication, and their knowledge builds on the insights of the modern human living conditions, their desires and behaviors and the outside world trends and tendencies. From these sources communicative concepts are created at a whole new level for each client focusing on experience, professionalism,

design, functionality and product advantage. The result of their work is experience concepts, which become part of the new product and a key part of the customer offers.

Website: www.stylt.se

Anagrama

They are a specialized brand development and positioning agency providing creative solutions for any type of project. Besides their history and experience with brand development, they are also experts in the design and development of objects, spaces and multimedia projects. They create the perfect balance between a design boutique, focusing on the development of creative pieces paying attention to the smallest of details, and a business consultancy providing solutions based on the analysis of tangible data to generate best fit applications.

Website: www.anagrama.com

Mr. Important

Mr. Important Design is an interior design firm that brings an ebullient perspective to interiors. Specializing in nightclubs, restaurants, hotels, bars and lounges, Mr. Important Design works closely with clients to achieve interiors that exceed expectations. Exuberant interiors that are designed to be remembered and talked about.

Collaboration with emerging designers and artists world-wide keeps the work fresh and surprising. Use of cutting edge technologies in lighting and materials coupled with a deep background in traditional furniture and decor help produce the unexpected thrill of these spaces.

Embracing many time periods, influences and esthetics, Mr. Important interiors are styled like hip-hop artists sample sound. Everything is available and usable, and the sparks fly best when kitsch and glamour, new and old, high and low all rub up against each other.

A depth of experience in designing, building and owning nightclubs, restaurants and bars gives the company a firm grasp on the realities clients face in creating an interior concept that can compete within a market saturated with concept. Mr. Important Design is there to help clients differentiate their brand experience into one that impresses patrons with indelible memories.

Founded in 2005 by Charles Doell, the interiors of Mr. Important Design have been widely featured in the international design press. Wallpaper Magazine said it best—Charles Doell is a 'design wiz-kid'.

Website: www.misterimportant.com

Rafael de Cárdenas

Rafael de Cárdenas received his B.A. from The Rhode Island School of Design and, following graduation, took a job at Calvin Klein, working for three years as a designer for the men' s collection. In 1999, he began pursuing an architecture degree at Columbia University, later transferring to UCLA where he received his Masters in architecture in 2002. His first project following graduation was working with the architect Greg Lynn on the redesign of the World Trade Center site. Their submission, a series of five buildings interconnected to create a cathedral-like space, was one of the six final entries. De Cárdenas then began work in the New York offices of special effects production house Imaginary Forces. As a creative director working on experience design projects, he oversaw a range of innovative concepts including the BMW Experience at their headquarters in Munich, and the HBO store in New York. In 2005, de Cárdenas opened his own design firm out of an office in New York's Chinatown. His interest in creating environments with moods, as opposed to any specific style, has allowed him to work with an array of clients. Using color, light, and pattern, de Cardenas has created artful, imaginative interiors for boutiques, restaurants and private residences in London, Rome, Athens, Chicago, Miami, New York, and The Hamptons. His work has been featured in Elle Dacor, Vogue Paris, The New York Times, Surface and Metropolis in addition to others.

Website: architectureatlarge.com

Outofstock

The design collective was born out of a fortuitous meeting in Stockholm, hence the name Outofstock. Gabriel Tan and Wendy Chua from Singapore, Sebastion Alberdi from Spain and Gustavo Maggio from Argentina met at Electrolux Design Lab 2005. They began their collaboration in 2006 and what started out as a creative experiment eventually became a long-term design partnership and the studio now works across furniture and interior design that counts Ligne Roset, Bolia, Foundry and FiftyThree as its clients. As a team comprised of culturally diverse individuals, Outofstock's creative energy derives from the differences of its members and not from any fixed formulas. What binds them is the vision of creating designs with a formal coherence that are relevant to the user.Outofstock is a registered business in Singapore with studio locations in Singapore, Barcelona and Buenos Aires.

Website: www.outofstockdesign.com

UXUS

Founded in Amsterdam in 2003, UXUS is an independent award wining design consultancy specializing in strategic design solutions for Retail, Communication, Hospitality, Architecture and Interiors. UXUS creates 'Brand Poetry', fusing together art and design, and creating new brand experiences for its clients worldwide. We define 'Brand Poetry' as an artistic solution for commercial needs. Artistic solutions target emotions; emotions connect people in a meaningful way. Design gives function, art gives meaning, poetry expresses the essence.

Website: www.uxusdesign.com

Claudia Meier

Website: www.claudiameier.ch

DarkDesignGroup

Website: www.darkdesign.ru

design spirits co., ltd.\Yuhkichi Kawai

Website: www.design-spirits.com

Jaime Hayon

Website: www.hayonstudio.com

Office dA

Website: www.officeda.com

Nendo

Website: www.nendo.jp

Design Research Studio

Website: www.designresearchstudio.net

Gwenael Nicolas \ CURIOSITY

Website: www.curiosity.jp

Acknowledgements

We would like to thank all the architects and designers for their kind permission to publish their works, as well as all the photographers who have generously granted us the rights to use their images. We are also very grateful to many other people whose names do not appear on the credits but who made specific contributions and support. Without them, we would not be able to share these beautiful commercial spaces with readers around the world.

Future Edition

If you would like to contribute to the next edition, please email us your details to: designbook@yahoo.cn